T0184970

The Profession and Practice of Technical Communication

This practical text offers a research-based account of the technical communication profession and its practice, outlining emergent touchpoints of this fast-changing field while highlighting its diversity.

Through research on the history and globalization of technical communication and up-to-date industry analysis, including first-hand narratives from industry practitioners, this book brings together common threads through the industry, suggests future trends, and points toward strategic routes for development. Vignettes from the workplace and examples of industry practice provide tangible insights into the different paths and realities of the field, furnishing readers with a range of entry routes and potential career sectors, workplace communities, daily activities, and futures. This approach is central to helping readers understand the diverse competencies of technical communicators in the modern, globalized economy.

The Profession and Practice of Technical Communication provides essential guidance for students, early professionals, and lateral entrants to the profession and can be used as a textbook for technical communication courses.

Yvonne Cleary is Head of Technical Communication and Instructional Design at the University of Limerick, in Limerick, Ireland. Her research interests include professional issues in technical communication, technical communication pedagogy, virtual teams, and international technical communication.

ATTW Book Series in Technical and Professional Communication

Tharon Howard, Series Editor

Communicating Project Management
Benjamin Lauren

Citizenship and Advocacy in Technical Communication
Godwin Y. Agboka and Natalia Mateeva

Rhetorical Work in Emergency Medical Services: Communicating in the Unpredictable Workplace
Elizabeth L. Angeli

Involving the Audience: A Rhetorical Perspective on Using Social Media to Improve Websites
Lee-Ann Kastman Breuch

Creating Intelligent Content with Lightweight DITA
Carlos Evia

Editing in the Modern Classroom
Michael J. Albers and Suzan Flanagan

Translation and Localization: A Guide for Technical Communicators
Bruce Maylath and Kirk St.Amant

Technical Communication After the Social Justice Turn: Building Coalitions for Action
Rebecca Walton, Kristen R. Moore, and Natasha N. Jones

Content Strategy in Technical Communication
Guiseppe Getto, Jack T. Labriola, and Sheryl Ruszkiewicz

Teaching Content Management in Technical and Professional Communication
Tracy Bridgeford

The Profession and Practice of Technical Communication
Yvonne Cleary

For additional information on this series, please visit www.routledge.com/ATTW-Series-in-Technical-and-Professional-Communication/book-series/ATTW. For information on other Routledge titles, visit www.routledge.com.

The Profession and Practice of Technical Communication

Yvonne Cleary

Routledge
Taylor & Francis Group

NEW YORK AND LONDON

First published 2022
by Routledge
605 Third Avenue, New York, NY 10158, USA

and by Routledge
2 Park Square, Milton Park, Abingdon, Oxon, OX14 4RN, UK

Routledge is an imprint of the Taylor & Francis Group, an informa business.

© 2022 Yvonne Cleary

The right of Yvonne Cleary to be identified as author of this work has
been asserted in accordance with sections 77 and 78 of the Copyright,
Designs and Patents Act 1988.

All rights reserved. No part of this book may be reprinted or repro-
duced or utilised in any form or by any electronic, mechanical, or other
means, now known or hereafter invented, including photocopying and
recording, or in any information storage or retrieval system, without
permission in writing from the publishers.

Trademark notice: Product or corporate names may be trademarks or
registered trademarks, and are used only for identification and
explanation without intent to infringe.

Library of Congress Cataloging-in-Publication Data
Names: Cleary, Yvonne, author.
Title: The profession and practice of technical communication / Yvonne
Cleary.
Description: New York, NY : Routledge, 2021. | Includes bibliographical
references and index.
Identifiers: LCCN 2021001537 (print) | LCCN 2021001538 (ebook) | ISBN
9780367558093 (hardback) | ISBN 9780367557379 (paperback) | ISBN
9781003095255 (ebook)
Subjects: LCSH: Communication of technical information. | Communication of
technical information--Study and teaching. | Technical
writing--Vocational guidance.
Classification: LCC T10.5 .C53 2021 (print) | LCC T10.5 (ebook) | DDC
601/.4--dc23
LC record available at https://lccn.loc.gov/2021001537
LC ebook record available at https://lccn.loc.gov/2021001538

ISBN: 978-0-367-55809-3 (hbk)
ISBN: 978-0-367-55737-9 (pbk)
ISBN: 978-1-003-09525-5 (ebk)

Typeset in Bembo
by SPi Global, India

Contents

List of Figures   *xi*
List of Tables   *xii*
List of Boxes   *xiii*
Series Editor Foreword   *xiv*
Acknowledgments   *xvi*

PART I

The Profession of Technical Communication   **1**

1   The Nature of Technical Communication   **3**

Introduction   3
Boundaries of Technical Communication   5
　Distinctiveness in Technical Communication   6
　Working Definition   8
Historical Roots   9
Theoretical and Research Bases of Technical Communication   9
　Theories in Technical Communication   9
　Research Base of Technical Communication   12
　Research Challenges   13
Technical Communication and Related Roles   14
　Job Titles in Technical Communication   14
　Synergies with Other Roles and Professions   15
This Book's Perspective   18
　The Scope of this Book   18
　Who Should Read this Book?   19
Summary of Chapter 1   20
Discussion Questions   21
References   21

**2   Technical Communication Education and Professional
     Development**   **24**

Introduction   24
Development of Technical Communication Education   25

Practicing Without Academic Qualifications   25
The Value of Education for the Profession   28
International Education and Training Programs   29
Finding and Choosing a Study Program   29
Program Titles   29
Academic Degrees: Certificate, Master's, and PhD   31
Choosing a Program: A Vignette   33
Typical Program Content   35
Examples of Education Paths   36
Program Spotlight: Communication and Media Management at the University of
Applied Sciences, Karlsruhe, Germany   37
Alternatives and Supplements to Formal Education   41
Examples of Training and Professional Development   43
Learning at Work   44
Certification in Technical Communication   45
The STC Certification Model   46
The Tekom Europe Certification Model   46
Certification Limitations   47
Education and Training Projects in Technical Communication   47
Technical Communication Body of Knowledge   47
The TecCOMFrame Project   48
Summary of Chapter 2   50
Discussion Questions   51
References   52

3   Technical Communication Communities   **54**

Introduction   54
What is a Community?   55
Professional Identity and Communities   55
Professional Associations   57
Features and Benefits of Professional Associations   57
Professional Associations in Technical Communication   59
Online Communities in Technical Communication   67
Communication Channels   68
Practitioner Communities   69
Volunteer Communities   70
Academic Communities   71
Academic/Practitioner Communities   71
Workplace Communities   73
Team Configurations in Technical Communication   74
Sociability at Work   76
Challenges and Limitations of Communities   77
Summary of Chapter 3   79
Discussion Questions   79
References   80

PART II

The Practice of Technical Communication   83

4   Technical Communication Activities, Tools, Genres, and Artifacts   **85**

Introduction   85

Competencies and Skills   86

  Research and Projects about Competencies and Skills   86

  Frequently Cited Competencies and Skills   87

  Transferable Skills   89

Core Activities   90

  Communication   90

  Information / Content Development   92

  Technological Dexterity   98

  Management   98

  Audience Analysis   100

  Content Strategy and Information Architecture   101

  Domain Knowledge   102

Tools and Technical Skills   103

  Range and Variety of Tools   103

  Common Tools and their Functions   105

  Other Technical Competencies   107

  Tools are just Tools!   108

Vignette: A Day in the Life of a Technical Communicator   109

Technical Communication Genres   112

  Purposes of Genre   113

  User Contexts and Genre   113

  Genre Conventions   114

  Typical Genres in Technical Communication   115

Common Technical Communication Artifacts   116

  Publishing Formats   116

  Manuals / User Guides   117

  Topic-based Formats   117

  Embedded User Assistance   118

  Training and E-Learning Content   118

  Websites   119

  Video and Animation   119

  Social Media Content   120

  Podcasts   121

Summary of Chapter 4   121

Discussion Questions   122

References   122

5   Technical Communication Workplaces   **125**

Introduction   125

Workplace Cultures, Sectors, and Environments   126

Workplace Cultures in Technical Communication   126
Industry Sectors in Technical Communication   128
Work Environments in Technical Communication   131
Vignette 1: The Content Writing Team Member   132
Global and Globalized Features of Contemporary Work   133
Remote Working   133
Vignette 2: The Contractor who Works from Home   136
Virtual Teams   137
Organizational Practices that Impact Technical Communication   138
Vignette 3: The Lone Writer   141
Professional Outlook for Technical Communication   142
Job Satisfaction in Technical Communication   142
Factors that Cause Dissatisfaction   144
Labor Market Trends in Technical Communication   145
Summary of Chapter 5   149
Discussion Questions   150
References   150

6   Technical Communication Futures   153
Introduction   153
The Future of Work and Professions   154
The End of an Era   155
The Implications of "The End of an Era" for Technical Communication   155
Transformation by Technology   156
The Implications of "Transformation by Technology" for Technical Communication   157
Emerging Skills and Competencies   159
*The Implications of "Emerging Skills and Competencies" for Technical
    Communication   159*
Professional Work Reconfigured   161
The Implications of "Professional Work Reconfigured" for Technical Communication   161
New Labor Models   162
The Implications of "New Labor Models" for Technical Communication   163
More Options for Recipients   163
The Implications of "More Options for Recipients" for Technical Communication   164
Preoccupations of Professional Firms   164
*The Implications of "Preoccupations of Professional Firms" for Technical
    Communication   165*
Demystification   166
The Implications of "Demystification" for Technical Communication   166
The Implications of Covid-19 for the Future of Work   166
The Implications of Covid-19 for the Future of Technical Communication   168
Durable and New Competencies   169
Durable Competencies in Technical Communication   169
New Competencies in Technical Communication   170
How Technical Communicators are Preparing for the Future   177
Summary of Chapter 6   179
Discussion Questions   180
References   180

PART III
Theories and Methods   **183**

7   Theories of Professions and Practice   **185**
Introduction   185
Profession and Practice: A Broad Perspective   186
  Defining the Field and Habitus   186
  The Importance of "Capital"   187
Professions and Professionalization   188
  What are Professions?   188
  Becoming a Profession   189
  Technical Communication Professionalization   189
  Problems with the Appeal to Professionalization for Technical Communication   191
Alternative Approaches to Studying Occupations   191
  The Symbolic Interactionist Perspective   192
  Professionalism and Professional Identity   193
Outcomes of Professionalization for Technical Communication   195
The Practice Turn in Theory and Research   196
  What is Practice?   197
  Communities of Practice   197
  Variation in Practice   198
  Researching Practice   199
Practice Studies in Technical Communication   200
  Variation in Technical Communication Practice   200
  Practice Narratives in Technical Communication   200
  Research Studies of Workplaces and Practices   202
  Education and Practice Intersections   203
Summary of Chapter 7   204
Discussion Questions   205
References   205

8   Data Gathering and Analysis   **210**
Introduction   210
Background: Why I Conducted this Study   211
  My Doctoral Study   212
  Taking a Global Perspective   212
Phase 1: Planning the Study   213
  The Research Approach   213
  Situational Analysis   214
  The Research Questions   217
Phase 2: Gathering the Data   218
  Literature Review   218
  Practice Narratives   219
  Content Analyses   221
Phase 3: Analyzing and Reporting the Practice Narratives   222
  Qualitative Analysis: Categorizing and Coding the Practice Narratives   222
  Writing the Vignettes   223

Limitations of this Project   224
Summary of Chapter 8   225
Discussion Questions   225
References   225

9   Conclusions and Future Directions   **227**
Introduction   227
Insights about the Profession and Practice of Technical Communication   227
Persistent Concerns: Strategic and Research Directions   228
   Strengthening our Professional Identity   230
   Research of/with/within International Contexts   231
   Practice- and Practitioner-focused Engagement   232
References   233

Appendices
Appendix A: Profession and Practice Resources   *234*
Appendix B: Ethics Committee Documentation   *238*

Index   *242*

Figures

1.1	Respondent locations globally	5
2.1	Disciplinary results for "technical communication"	31
2.2	Tekom Europe, The TecCOMFrame Academic Competence Framework	49
2.3	A prototype curriculum at the master's level	50
4.1	Markdown editor and preview example	108
4.2	Microsoft Word™ sample templates	114
5.1	A plaque at the University of Limerick expresses the institution's commitment to learning through a motto on the crest	127
5.2	Bureau of Labor Statistics, US Department of Labor, *Occupational Outlook Handbook,* Technical Writers	146
5.3	The results of a search for "content" jobs on IT Jobs Watch	149
7.1	Tina the Technical Writer	201
8.1	Situational analysis of the technical communication profession and practice (V1: October 2019)	216

Tables

1.1	Summary of purpose, from the technical communicator's and the user's perspectives	8
1.2	The development of the technical communication profession in the past 100 years	10
2.1	Time line of educational developments in technical communication	26
2.2	International programs and professional development in technical communication	30
2.3	Structure of the bachelor of arts (BA) in Communication and Media Management	39
2.4	Master of science (MSc) in Communication and Media Management: Modules	40
3.1	Selected professional associations in technical communication: Practice-focused	60
3.2	Selected professional associations in technical communication: Academic-focused	62
3.3	Conferences in technical communication	64
4.1	Competencies and skills required in technical communication	88
4.2	Tools commonly used in technical communication	106
5.1	Regions and work sectors	130
5.2	Job titles matrix	147
5.3	Jobs advertised by title and recruitment site (September 28, 2020)	148
6.1	New competencies: Summary of practice narrative responses	170

Boxes

1.1 You should read this book if...   18
3.1 Practice narrative excerpts about the purpose of membership
     in professional association   66
3.2 Practice narrative respondents describe how they use LinkedIn
     and Twitter   68
4.1 When to use video, screencasts, or animation   120
6.1 Practice narrative excerpts about perceptions of adaptability   170
7.1 Overview of professional traits   188

Series Editor Foreword to
The Profession and Practice of Technical Communication by Yvonne Cleary

Over the past five years, the Association of Teachers of Technical Writing (ATTW) Series in Technical and Professional Communication has been engaged in a constant search for authors and titles that can produce and deliver scholarship which will serve as long-term resources for both practitioners and students of our profession, and Yvonne Cleary's work, *The Profession and Practice of Technical Communication,* is exactly the sort of broadly focused, internationally sensitive, and meticulously researched compendium on professional and technical communication that our field has needed for some time now. Whether you're a teacher looking to introduce your students to careers in technical and professional communication or a practitioner looking for resources you can use to keep pace with future trends and strategic routes to the development of technical communication in industry, this is a book you'll want to have on your bookshelf. Indeed, as one of the external reviewers for the book put it, "The status and history of technical communication has long been a major topic for technical communication researchers. *Industry Perspectives on the Profession and Practice of Technical Communication* promises to be one of the most significant books ever published on the topic."

The Profession and Practice of Technical Communication offers readers a research-based account of the technical communication profession and its practice, from education and careers, through day-to-day activities, to possible future developments. The text is based on the analysis of primary narrative data from industry practitioners, together with analysis of academic research and online resources. The book includes examples of industry practice and workplace vignettes that describe practitioners' experiences when working in technical communication. If you're a member of a teaching faculty like myself, you know how difficult it can be to actually *show* students credible examples of the kind of writing and work environments for which our classes are intended to prepare them. As a teacher, I can "tell" my students about professional practices they can expect to experience, but Yvonne's vignettes "show" them and transcend the walls of the academy by bringing the voices of global practitioners into the classroom. These examples and vignettes suggest a range of entry routes and possible careers, sectors, workplaces, communities, processes, activities, and envisioned futures. This approach is central to helping readers understand what technical communicators do in the modern, international economy.

Technical communicators operate in globalized environments in roles and workplaces that are diverse and dynamic. Because the profession has undergone vast changes due to technology and globalization, Dr. Cleary's work here seeks to identify commonalities across practice and to explore the distinctive features that currently define the profession. She identifies essential characteristics of the technical communication profession and its practice, through insights from industry-based practitioners in 13 countries as well as from academic research, trade publications, and online sources. This content will be valuable for current students, practitioners and new entrants to the profession, and technical communication instructors and researchers. The vignettes and excerpts from the primary data which Yvonne has compiled will enable teachers, students, and practicing professionals to analyze and decode various aspects of technical communication, including educational routes, workplaces, and typical activities. The book includes an appendix of online resources that can serve course designers, researchers, and students by encouraging further reading and research and by connecting students to international technical communication resources and communities.

Again, I'm thrilled that this book is part of the ATTW Series in Technical and Professional Communication. The topic is totally consistent with and wholly appropriate for the Series and provides exactly the blend of solid academic scholarship and industry application that we wish to provide to readers of the Series.

Tharon W. Howard
Editor, ATTW Series in Technical and Professional Communication
December 1, 2020

Acknowledgments

Sincere thanks above all to the 62 technical communicators who wrote narrative accounts of their education, work histories, daily work, and/or future career expectations. It is truly the case that this book would not have been possible without those vivid and generous contributions. Sissi Closs agreed without hesitation to be interviewed and for her master's program to be profiled for Chapter 2. She was generous with her time and ideas, and I am certain that her experiences will help teachers who develop technical communication programs and students who take those programs.

This book builds on the work of many technical communication scholars, and I am indebted to their insights. I am also indebted to collaborators in industry and academia with whom I have had lively discussions about the profession and practice of technical communication over the past decade. Thanks especially to the members of the TecCOMFrame project team for igniting my interest in the profession in different European countries: Sissi Closs, Zygmund Drazek, Jan Engberg, Voichita Ghenghea, Joyce Karreman, Birgitta Meex, Patricia Minacori, Julia Müller, Anke Neytchev, and Daniela Straub.

I was fortunate to be awarded sabbatical leave by the University of Limerick in 2019-2020. I owe particular thanks to the Dean of the Faculty of Arts, Humanities and Social Sciences, Prof. Helen Kelly-Holmes, and the Head of the School of English, Irish, and Communication, Prof. Michael Griffin, for supporting the sabbatical application and later for enabling me to prioritize the book when the sabbatical leave was cancelled due to the pandemic. I thank Rosemary Fogarty for helping me to plan the sabbatical and Prof. Meg Harper, former Head of the School of English, Irish, and Communication, for encouraging me to apply. My colleagues in Technical Communication and Instructional Design were also accommodating: I am grateful to Darina Slattery, Ann Marcus-Quinn, and especially Elaine Walsh and Margaret Grene, who undertook my teaching and master's supervision while I was on leave. Margaret Grene also read a draft of the manuscript and I very much appreciated her keen editorial eye.

Tharon Howard has been kind, interested, enthusiastic, generous with his time, quick to respond, and just very helpful since I first suggested this topic to him and throughout the process. His feedback on my first proposal shaped the direction I took with this book. Kirk St. Amant read an early draft and provided many insightful comments and suggestions that also shaped my thinking and that enabled me to write a much sharper and more targeted proposal. Three reviewers gave feedback on the manuscript proposal in a level of detail that impressed and terrified me. Those reviews certainly helped to make this a more reader-friendly and useful work. Brian Eschrich at Routledge had faith in the project, and Grant Schatzman was exceptionally attentive and responsive, providing quick, clear, and detailed answers to all my questions. I am also very grateful to Stephen Poole for his meticulous copyediting.

For permission to reproduce screenshots, I am indebted to tekom Europe, IT Jobs Watch, and Studyportals/Mastersportal for excerpts from their websites; the Bureau of Labor Statistics for use of an excerpt from the *Occupational Outlook Handbook*; and Microsoft for a screenshot from the Microsoft Word™ application. I am indebted to Andrews McMeel Syndication for permission to use the Tina the Technical Writer strip in Chapter 7.

Friends and colleagues have supplied endless help, friendship, and solidarity. In no particular order, thank you to Madelyn Flammia, Caoilfhionn Ni Bheacháin, Jean Conacher, Barbara Geraghty, Claire Moloney, Philip Dowdall, Fergal Quinn, Rosie Gowran, and Anne Baragry. My parents, siblings, and their families are kind, generous, and interested, above and beyond the call of duty: thank you to Mam, Dad, Carmel, Ann, Paul, Neil, Mark, Luke, Mike, Lisa, Eoin, and Colin. And last, but of course not least, thanks to Michael, who has been patient, kind, good-humored, and fun to be around while I have worried away at this project.

Part I

The Profession of Technical Communication

This book is organized into three parts:

Part I: The Profession
Part II: The Practice
Part III: Theory and Methods

The first of the book's three major sections, Part I explores the nature of the technical communication profession, education and training opportunities, and the communities and professional organizations that support technical communicators in their work. This part has three chapters:

- Chapter 1 offers a working definition and provides background information about the profession.
- Chapter 2 discusses educational and professional development opportunities and projects in technical communication.
- Chapter 3 explores the communities that support technical communication practitioners.

Whether you are a student, a practitioner, or a teacher, this part of the book will help you to recognize key professional issues of concern in technical communication: the distinctive character of our work, routes into the profession, and communities that support and sustain us.

1 The Nature of Technical Communication

Introduction

Some professions have a high profile in society, media, and popular culture. Most of us have gone to school, so we can describe at least some of the day-to-day activities of a teacher. From our transactions in shops and financial institutions, we can imagine the work of a shop assistant or bank clerk. Even those of us who have never been to hospital have seen enough television shows about physicians and surgeons to be able to name some of their activities, however sketchily. The same is true for the legal professions; although we may not have had a brush with the law, most of us have absorbed the procedures and language ("Objection!" "Overruled!") from films and courtroom dramas.

By contrast, technical communication, like many newer professions, has a lower profile in the public consciousness. Its activities are often unknown to outsiders. If you already study or work in technical communication, you probably recognize Kirk St.Amant's (p. 1)[1] descrip-

This chapter explores both the variety and the common threads that make technical communication diverse but distinctive and recognizable.

We begin with a discussion of the **boundaries of technical communication** to arrive at a **working definition**.

An overview of the **historical roots and theoretical and research bases** of technical communication will help you to see how your profession developed and which theories and research studies are reflected in your practice.

Technical communication encompasses a broad range of possible careers and **job titles**. This profession has connections to a constellation of related professions. You will identify the intersections with other disciplines, including content strategy, localization and translation, instructional design, and user experience (UX). These intersections are evident throughout the book.

The chapter concludes with a description of **who should read this book**.

tion of how people respond when you tell them your job title or study program:

> Tell someone that you work in the area of 'technical communication,' and chances are you'll receive a curious, somewhat confused look in response. It doesn't matter if you're an academic researcher or an industry practitioner; the look is almost always the same. And we've all seen it at some point or another in our careers—it's that expression of 'I kind of know what that is/ what you mean, but I'm not 100% sure.'

Because of the variety of activities and workplaces, even those of us who study, practice, or teach technical communication are unfamiliar with some aspects of the profession and practice.

This book evolved from my teaching experiences and research interests. I have taught a course about the technical communication profession for over 15 years. During that time, I have struggled to find a book that explains what it is like to work in technical communication, that covers industry perspectives, and that discusses common activities and typical workplaces. Because I have worked in a university for most of my career, I have always relied on input from industry practitioners to help me to prepare students effectively for contemporary workplaces. I typically use three sources to develop courses that have a broader perspective and that balance practical and academic content:

- Academic work about industry practice: presentations, books, projects, articles, papers, and reports.
- Direct input from industry practitioners.
- Websites, blogs, social media, and other online resources developed by professional associations and individual practitioners.

This book uses the same range of sources to give you insight into contemporary technical communication in industry. Throughout this book, you will find descriptions and discussions based on my interpretation of what I read about technical communication, professions and professionalization, practice, and work more generally. The primary research involved a narrative survey of practitioners and had 62 responses from individuals in 13 countries in Europe, North America, and Australia.

The survey results, which I refer to as practice narratives, are included throughout the book. You will find **examples of industry practice** in the form of excerpts from these narratives and **vignettes** – composite stories based on the practice narratives – that describe practitioners' experiences of initiation into the profession and their practice in technical communication. I also analyzed academic and trade publications and online resources. (Chapter 8 discusses the research methods in detail.)

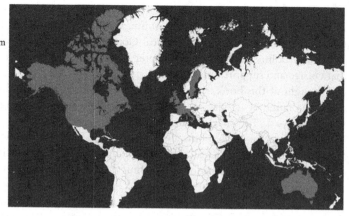

Ireland
France
United Kingdom
Italy
Hungary
Sweden
United States
Belgium
Austria
Switzerland
Canada
Australia
Israel

Figure 1.1 Respondent locations globally.

I gathered the narrative data in late 2019, shortly before the Covid-19 pandemic changed how so many of us live and work throughout the world. Although the practitioners who responded to the survey obviously did not foresee the pandemic, I analyzed several online sources that provide insight into its potential impact for technical communicators.

Boundaries of Technical Communication

Technical communication is related to and "borrows from" many other fields of study, including "design communication, speech communication, and rhetoric and composition as well as … psychology, education, and computer science" (p. 175).[2] It is a profession of diverse practices, where practitioners have many different job titles, participate in multiple communities, and operate in many types of work setting. Because of role diversity, we also have immense scope to specialize and plentiful opportunities in the labor market. Later in this chapter, you will examine job titles that all fall under the umbrella of technical communication. You will also look at related roles in neighboring disciplines.

Individuals come to this profession from many academic backgrounds (see Chapter 2 for a discussion of education and training opportunities). It is not uncommon for technical communication to be a second career, and we often have degrees in disciplines other than technical communication.[3] This situation has led to a need for, and supply of, professional development opportunities. Technical communicators are on a continual learning journey. As one practice narrative respondent explained:

> Since I was never formally trained as a technical writer, it was a trial-and-error approach for me. I'd say I am still learning every day, which is what I love most about my job.

Several professional associations and communities support our work. These include the Society for Technical Communication (STC), tekom Europe, and the Institute of Scientific and Technical Communicators. We are also involved with many online communities. (Chapter 3 discusses the range of communities that engage and support our work.)

Throughout this book, role diversity is a prominent theme. Chapter 4 discusses the range of activities we undertake, artifacts we develop, and genres in which we work. Technical communicators can work in any economic sector. Typical sectors are private businesses and corporations, government agencies and institutions, professional associations and societies, and education and nonprofit institutions.[4] In addition, technical communication service or consultancy firms undertake outsourced work. Respondents mentioned these sectors in the practice narratives: information and communications technology (ICT) (software, information technology (IT) services, and internet applications), medical devices, biomedical, financial services, research, and higher education. The ICT sector was by far the most prominent one mentioned in the narratives. (Chapter 5 explores occupational sectors and workplaces in more detail.)

It seems likely that the diversity and variety that characterize our work, workplaces, educational offerings, and communities will continue into the future. (See Chapter 6 for some possible future directions for our work.)

If we are so diverse, what, if anything, makes us distinctive? How do we know a technical communicator from another professional, such as a content strategist or an instructional designer? If our job title is not "technical writer" or "technical communicator," how do we identify with this profession?

Distinctiveness in Technical Communication

In some professions, practitioners are recognizable by obvious markers: their titles or degrees, the physical activities they perform, or the uniforms they wear.[5] For example, in some hospitals, the color of the scrubs worn by medical staff indicates their specialism or level of seniority. In technical communication, these markers are less obvious, but many features distinguish us from other professionals. Exploring these commonalties helps us to identify with our profession.

Recognizing distinctiveness is important because it enables us to align ourselves to a profession, train appropriately, have career mobility, envisage a career trajectory, join communities that can support us, develop our practice, and strategize for our future careers and for our profession.

The large professional associations define technical communication according to the industry focus of the role. The STC[6] sees **three strands to the scope of the profession**:

- Communicating about technical or specialized subjects.
- Communicating through technology.
- Providing instructions for procedural tasks of any kind.

Tekom Europe[7] has a concise definition that focuses on **process and products**:

> Technical communication is the process of defining, creating and delivering information products for the safe, efficient and effective use of products (technical systems, software, services).

The Australian Society for Technical Communication[8] defines technical communication according to **products**:

> Technical information in any format, including printed, electronic, mobile, online help.

The Technical Communication Body of Knowledge,[9] an initiative of the STC (and discussed in more detail in Chapter 2), provides a definition that **prioritizes information users**.

> The delivery of clear, consistent, and factual information—often stemming from complex concepts—for safe and efficient use and effective comprehension by users. Technical communication is a user-centered approach for providing the right information, in the right way, at the right time so that the user's life is more productive.

Karen Schriver (p. 277)[10] also defined technical communication in terms of its **value for information users**.

> Well-written and well-designed artifacts can promote comprehension of ideas in science, education, and technology, enabling people to build coherent representations of complex content, such as understanding nanotechnology or macroeconomic theory.

She distinguished writing "as a profession" from writing "as a work tool." Technical communicators write – or, more broadly, produce content – as their profession. Although many people develop content as *part* of their job, this book's focus is anyone whose *main role in their job* concerns technical communication.

Working Definition

As we see from the definitions, technical communication involves communicating, explaining, and helping people to learn, understand, use, and interact with specialized concepts or technologies. Technical communicators design interventions to help people to use technology, services, or information, safely and effectively. These interventions take many forms, from print content to knowledge bases, developer documentation, e-learning courses, video, and myriad digital formats. This is both a humanistic and a technical discipline because of its focus on enabling people to interact with devices, services, and content. Therefore, it is at the heart of **digital transformation**. Technical communication is an inclusive profession, and it **encompasses many job titles**.

> In the absence of another agreed term, I have used "technical communication" throughout this book as shorthand for the many jobs and roles where individuals develop content that helps people to use technology, services, or information.

The definitions of technical communication show that it can be described from the perspective of professionals who develop content and from the perspective of users of that content. These perspectives are slightly different. Table 1.1 summarizes what the definitions tell us about how professionals practice, and what users need from, technical communication.

Table 1.1 Summary of purpose, from the technical communicator's and the user's perspectives

As a professional, I ...	As a user, I need ...
Develop information products.	Usable and user-centered information products.
Write clear and safe procedural instructions.	Instructions that enable me to follow a procedure safely and correctly.
Communicate clearly, precisely, and concisely about technology, technical concepts, or specialized topics.	Content that explains technology, technical concepts, or specialized topics in ways I can understand.
Explain concepts.	To understand and apply concepts.
Provide information when and how the user needs it.	To be able to access the information I need, when I need it.
Produce content that enables users to understand devices, concepts, and procedures and to apply their understanding safely.	Content that is easy to understand and use and that prioritizes safety.
Select appropriate tools and approaches for user-centered and responsive content.	Information in the format and mode I need, suitable for my device and context.
Develop inclusive and accessible content.	Content that I can access, use, and understand regardless of my personal or contextual restrictions.

This book focuses on your profession and your work or your future work. Because technical communication is a user-centered profession, however, it is important to understand the user's perspective and how it affects your job. For this reason, we will return to the user-centered focus throughout the book.

Historical Roots

Knowing about the history of technical communication helps us to understand why the profession and practice evolved in the ways they did, why we currently work the way we do, and how the profession might continue to change in the future. As Teresa Kynell explained, "history forms the foundation on which the future will be built" (p. 143).[11]

The time line in Table 1.2 provides a snapshot of key phases in the development of the profession of technical communication and the impact of these developments on practice. Although technical communication existed in support of science and industry for centuries, it developed into a mainstream occupation in the twentieth century.[12]

As the time line shows, technical communication has changed continually as new technologies have emerged. The pace of technology change is not slowing, so you can expect that your profession will continue to be defined by change into the future. Chapter 6 discusses future developments in more detail.

Theoretical and Research Bases of Technical Communication

When you work in industry, you need to be able to explain the theoretical and research foundations for your work. This foundation gives you confidence in your work, and credibility with your colleagues, because you can justify your decisions and show that your work is specialized and methodical.

Theories in Technical Communication

Although technical communication is quite practical, we base our work upon theories from many disciplines.[19]

A theory is a proposition that describes a relationship among variables.[20] In doing so, it explains a situation or event. We test theories in our research, and theory also influences our practice.

Technical communication is grounded in theories from disparate disciplines (e.g., psychology, rhetoric and composition, sociology, communication, media, education, computer science, and design). For example, in the four issues

Table 1.2 The development of the technical communication profession in the past 100 years

Date	Development	Consequences
Pre–20th century	Although instructions were needed for technical and mechanical products and inventions, technical communication was not a professional activity.	Engineers or other professionals wrote instructions. Technical communication did not exist as a specialized field of work.
1910s	Writing courses were developed for engineers.	Technical writing as an academic discipline emerged in engineering programs. (See Chapter 2 for more information about the educational development of the profession.)
1920s and 1930s	These decades saw mass electrification and important consumer developments (e.g., the invention of the telephone, and mass production of cars).	Electrification and inventions increased the number of machines in use and the need for instructions to help lay users to operate the machines.
1940s	Military activities during and after the Second World War increased the need for technical communication.	"Defense-related production influenced the development of technical writing … as the sophistication of weaponry increased, manufacturers needed writers to explain that technology to workers who lacked a technical background" (p. 148).[11]
1950s	The US and many parts of Europe experienced a post-war economic boom.	Writers "who had been busy providing wartime technical information remained busy as wartime technologies were applied to peacetime uses" (p. 155).[13] Large technology firms such as General Motors and General Electric began to use the expertise of technical writers. New consumer goods also required instructions.[14]
1960s and 1970s	This period saw military, aerospace, and nuclear expansion.	These developments further increased the need for technical communication, in the US and also in other parts of the Western world. Nevertheless, technical writing often continued to be a secondary occupation for individuals with no specialized training. For example, in the UK, the task of writing technical documentation to support aviation and engineering products and procedures was often undertaken by engineers.[15]

1980s	The manufacture and widespread use of the personal computer led to a paradigm shift.	Lay users needed instructions to be able to use personal computers and software. This era saw the emergence of the modern technical communicator, whose job is to explain technology to users. John Carroll (p. 2)[16] described the development of early mass-produced desktop computers as a "revolution in technology to support human activity," but one that caused "nightmares" for users, who needed empathetic instructions to improve their experiences with technology.
1990s	The internet "began to percolate into mainstream society" (p. 5)[17] in the early 1990s, spawning the so-called "information age." The infrastructure was based on decades of technology research.	The widespread and fast adoption of internet technologies led to a move toward online delivery of modular content. For example, online help was packaged with software alongside, or instead of, print manuals. An abundance of online information led to a need for single sourcing and content management methodologies.
2000s	As internet technologies developed, more online services were enabled, and Web 2.0 "harnessed the collective intelligence available on the Web, either via software [...] or by exploiting the willingness of users to engage" (p. 17).[17]	User-generated content became prominent on wikis, online reviews, and discussion forums, for example. More self-service platforms and web-based services came into use, increasing work opportunities for technical communicators and increasing the need for topic-based formats. The publication and adoption of the Darwin Information Typing Architecture (DITA) standard led to more structured content and more short-form and topic-based formats. As a result, the traditional print manual has been in decline in the past two decades in most sectors. Mirel and Spilka noted the great strides that technical communication made in the twentieth century but they also predicted threats to "coherence, cohesion and unity" (p. 1)[18] from modular and personalized content.
2010s	Changes in this era have been driven by developments like the internet of things, artificial intelligence, and questions about the future of work as the potential of smart machines increases.	This decade saw increased scope for personalized and intelligent content. Embedded content and microcontent are becoming important delivery formats. Artificial intelligence and automation have had as yet only minor impact on how technical communication content is produced.

published in one year (2018), studies in the *Technical Communication* journal drew on theories from this range of disciplines:

- Rhetoric.
- Communication studies
- Instructional design.
- Engineering.
- Ethics and social justice.
- Intercultural communication.
- Computer science.
- Management/organizational behavior.
- Political science.

Being able to draw on an expansive theoretical background is both intimidating and exciting. It will be impossible to read everything you would like to, but your study program, and your job, will be challenging and intellectually stimulating. The variety of theoretical approaches is another reason why technical communication can be more challenging to describe and explain than disciplines that have theoretical bases in only one or a small number of areas. It may be both a cause and effect of the profession's diversity.

Research Base of Technical Communication

Theory and research are interdependent because research studies often aim to verify, develop, build upon, or refute theories. Theories inspire research studies and influence practice in technical communication. As a professional, you should be able to point to the research foundation on which your work is based since "the identity of any academic field is based in part on the research it conducts" (p. 175).[2]

In technical communication, several journals publish research that is directly relevant to our profession. These are the best known:

- *Journal of Business and Technical Communication.*
- *Technical Communication.*
- *IEEE Transactions on Professional Communication.*
- *Journal of Technical Writing and Communication.*
- *Technical Communication Quarterly.*
- *Computers and Composition.*

You will also find descriptions of useful and relevant research studies in journals in related fields like rhetoric and composition, education, instructional design, UX, computer science, media and communication studies, and management. In addition, several academic publishers publish books describing research in technical communication.

Using this global research question, Carolyn Rude described the research avenues in technical communication (p. 176):[2]

"How do texts (print, digital, multimedia; visual, verbal) and related communication practices mediate knowledge, values, and action in a variety of social and professional contexts?"

She argued that all technical communication research is concerned with some aspect of this question. She also identified four research strands:

- Disciplinarity: what technical communication is and how technical communicators identify with their profession.
- Pedagogy: how to design and teach technical communication programs and courses.
- Practice: how to create effective technical communication products.
- Social change: how technical communication texts change society.

Research is linked to professional identity,[21] and research studies can have a considerable impact on our profession and its practice. For example,

- Karen Schriver's work in her book, *Dynamics in Document Design,*[22] emphasized the value of user research in practice and gave technical communicators increased impetus to explore user behaviors, user contexts, and user interactions with texts. It also offered technical communicators a solid foundation for the decisions they made in their practice.
- Research about the profession in the 1990s and 2000s enabled technical communication to become more professionalized, particularly in the US. For example, many more academic programs were developed, and the profile of the job and the profession increased.
- Recent research about social justice has highlighted the need to develop content that is inclusive and respectful of all users. This research strand has also examined the experiences of marginalized individuals in technical communication workplaces.

Research Challenges

Although academics conduct research on topics that have workplace relevance and practitioners understand and value research, there is a divide between academic research and practice concerns in technical communication.[21,23] Most published research in the past 20 years has focused on pedagogy, rhetoric, and genre, topics that are likely of more interest to academics than to industry practitioners.[24] For this reason, academic research can seem distant from everyday work contexts. This is a challenge that academic researchers need to act upon. Our interdisciplinarity can also complicate our attempts to identify and distinguish technical communication research.[23] A further challenge is that published work about technical communication and its practice is usually conducted by North American researchers[24] and describes North American contexts. Even where research explores international contexts, it is often written by academics based in North America and consequently represents international situations from this sociocultural point of view. I hope this book takes one step toward addressing some of these research challenges and, in doing so, points toward additional avenues for collaboration among all stakeholders in our profession.

Technical Communication and Related Roles

Some of us are probably clear about how our work aligns unequivocally with the technical communication profession. Others of us may see ourselves on the periphery, either because we have a different job title or because we work in a related but distinct field.

Job Titles in Technical Communication

The profession of technical communication encompasses many job titles. "Technical writer" has been the most common job title since the 1980s. In the US, the Bureau of Labor Statistics has recognized "technical writer" as an official job since 2010.[25] This recognition is important for the profile of the profession. In Germany, too, the profession of "technischer Redakteur" (technical writer) is recognized by the Bundesagentur für Arbeit (federal employment agency). Indeed, "technical writer" is a recognized profession in several Western countries and regions with globalized economies, among them Spain, Australia and New Zealand, and Canada. In most countries, though, it is not listed in national occupation classifications.

Although "technical writer" continues to be widely used as a job title, many additional titles are used. Academic programs and professional associations began to use the "technical communicator" and "technical communication" labels in the 1990s. These were seen as more inclusive terms, indicating that the profession involves more than writing but encompasses several communication strategies. "Technical communicator" may be a successful umbrella label, and it has been adopted in many parts of the world as a professional title. It never really gained currency in the labor market, however.[26] A search for "technical communicator" on most job search tools returns few direct results.

Since the early 2000s, multiple new job titles have emerged. These titles often reflect the digital skills required and the increasing focus on "content" or "information" rather than documentation (a term that has connotations of print and long-form, instead of modular, content). The practice narrative respondents had a range of job titles. "Technical writer" was by far the most common; other titles included "instructional designer," "content strategist," "technical author," "information developer," and "information manager."

Multiple job titles are a result of both digitalization and diversification and the associated need to have many skills besides writing and communication.

This diversity has been evident for decades.[26] In 2002, Anschuetz and Rosenbaum (p. 250)[27] noted that diversification was leading to an increasing range of job titles. They suggested that the variety of titles could be a positive direction for the profession, signifying an expanding definition of technical communication.

> The transition to expanded roles often involves new job titles that do not immediately bring to mind traditional technical communication activities. However, title changes do not mean that individuals in these new roles cannot still feel a kinship with technical communication in spirit, allegiance, and perspective.

Job titles, sectors, and labor markets are discussed in more detail in Chapter 5.

Synergies with Other Roles and Professions

Although job titles are abundant, it is less clear how roles with different titles are both connected to and distinctive from each other. Do these different job titles have different skill requirements? One study found that job titles "sometimes give us clues as to the skill sets needed by technical communicators, but just as often, they do not" (p. 14).[26] Some titles are largely synonymous with technical communication (e.g., information development and user assistance). Others suggest very close connections or subdisciplines, like information design, content strategy, and UX writing. Still others are distinctive (e.g., translation) but nonetheless are closely linked to our profession.

Information design

Information design involves "the design, usability, and overall effectiveness of 'content put into form' – of verbal and visual messages shaped to meet the needs of particular audiences."[28] Information design, somewhat like "technical communication," is a label that does not transfer directly to the labor market. It tends to be a skill that is required for roles in technical communication, graphic design, web design, and other related disciplines, rather than a discrete job title.

Content strategy

Content strategy describes both a model and a process for treating content, usually within an organization, throughout its life cycle. Essentially, it is "the creation, publication, and governance of useful, usable content."[29] Content strategy is sometimes identified as a technical communication skill, but it is also a skill that is required in other disciplines (e.g., marketing), and it is also a distinct job title.

Instructional design

Instructional design incorporates theoretical and methodological approaches to the development and delivery of training materials, including online training programs. Technical communication and instructional design are synergistic disciplines. Technical communicators often develop training programs. Adopting theories and strategies from instructional design helps us to create engaging content that encourages active learning. At the University of Limerick, our master's program incorporates both technical communication and instructional design, and graduates have job opportunities in both these disciplines as well as in hybrid roles like developing online training materials.

User Experience (UX)

Since the 1990s, a focus has emerged on designing content and products with the end user's experience at the center of the process. The discipline of UX evolved from human-computer interaction and incorporates concepts from traditional usability as well as "beauty, hedonic, affective or experiential aspects of technology use" (p. 91).[30] UX prioritizes "be" goals, how you feel about a product or service, whereas usability is about "do" goals, what you can accomplish with a product or service.[31] Cory Lebson[32] sees UX as an umbrella term for several careers, including technical communication. If you work in technical communication, however, you might have the converse perspective (as I do): technical communication is the umbrella, and UX is an aspect of our work and a possible specialism. In any case, there is a lot of crossover in these fields, and you will see UX roles and activities discussed in several chapters of this book.

Translation and Localization

Translation and localization are linked professions that involve the preparation of content for different languages or cultures. Linguistic translation is one step. Localization involves the complementary, or alternate, step of adapting content for different cultures or different locales (or local markets) or both. This adaptation can include modifying graphics, color schemes, layouts, time, date and address formats, and other features. Although the skills needed for technical communication are distinct from those needed for translation and localization, in many countries it is common to join the technical communication profession from a translation background.[33] One practice narrative described how their work in translation led them to become interested in, and train to work in, technical communication.

> While working in translation, I had continuous contact with technical writers and their content. I became interested in the subject and went back to school to get a master's degree to do it myself.

Marketing/Corporate Communication

Marketing communications "refers to all activities – research, strategies, or tactics – that support the selling of products and services" (p. 271).[34] As the channels used to communicate with customers are increasingly online, marketing professionals need to be able to write concisely for social media and to use data analytics, search engine optimization, website and information design techniques, UX knowledge, and other technical communication skills to deliver effective marketing campaigns. These requirements have led to a natural crossover between roles in marketing communication and technical communication. (In an interview reported in Chapter 2, Professor Sissi Closs explains how graduates of the Communication and Media Management program at the University of Applied Sciences in Karlsruhe, Germany have opportunities to work in marketing communication roles.)

Engineering and Software Development

Although the professions of engineering and software development are quite different from technical communication, we have a lot of interaction with professionals in these disciplines, depending on the industry and domain we work in. You will read examples of these interactions in later chapters. We often write content about software, hardware, or IT applications, for example. Many of us rely on subject-matter experts from these disciplines to help us to ensure that our content is accurate. We sometimes use development and management frameworks similar to those of these professionals. For example, Agile methodologies are common in software development teams that include technical communicators. It is also common for us to use the same project management tools. Because of the high level of interaction, it is not unusual for individuals who work in these professions to move into careers in technical communication.

Other Professional Intersections

I have outlined some of the more common role intersections, but technical communication can permeate many roles and sectors. In addition, many technical communicators have moved into this profession from entirely different career paths. One practice narrative respondent explained how their experience of writing manuals and instructions in a health-care profession spawned a technical communication career:

> Although not my job title, one of the things I set about doing was writing product manuals and developing training materials. I had been employed to carry out training on a computer system, and to liaise between the software developers and the clinicians, and soon discovered they had no written materials at all. It was while working here that I switched the path of my first degree from health care to computer science, and that last course of that degree was the one that made me realise I was actually a technical author.

This Book's Perspective

The variety in this profession, of roles, titles, skills, and communities, makes our work interesting but also can make it more difficult for us to recognize what technical communication is and what technical communicators do. Our professional identity may be in danger of fragmenting.[35]

My intention in this book is to help **students and professionals who develop specialized content** to recognize the common themes of our work and the educational diversity, communities, activities, workplaces, and potential in this profession. The book focuses **on technical communication practice in industry**, but the content is also relevant for practice in academic, government, and nonprofit sectors. Indeed, there are many parallels and commonalities across the sectors.

Regardless of your job title or the sector you work in, the book's contents will be relevant to you if most or all of the following statements apply to your current or future career.

Box 1.1   You should read this book if …

- ☐ Your **main job** (or your career objective) is developing content.
- ☐ You develop content **to communicate about or explain technical or specialized topics, services, or products**.
- ☐ You **use technology to communicate**.
- ☐ The **audience** for the information you develop is end users, developers, or other stakeholders (e.g., educators, trainers, decision makers, or gatekeepers).
- ☐ You work in an international or **globalized** environment or both.

The book takes a **global perspective** on the technical communication profession and its practice. As an academic discipline, technical communication is perhaps more advanced in North America than in other regions of the globe. Therefore, most available research published in English discusses aspects of practice in this region and particularly in the US. Where possible, I have incorporated examples of international education, communities, research, and practice to increase our understanding of this profession throughout the world.

The Scope of this Book

The profession of technical communication has undergone vast changes due to technology and globalization, and our roles and workplaces continue to evolve. I hope that, in the midst of this change, this book will help you to identify with your profession. The book uses research findings to establish **commonalities across practice** and to explore some **distinctive features** that will help you to recognize and identify with your profession.

The book is organized into three parts, which explore the profession (Part I), the practice (Part II), and the theory and methods I used in the study (Part III). If you are new to technical communication, you probably have many questions about this profession and its practice. This book helps you to begin to answer these questions, and it provides resources so that you can answer additional questions yourself.

- What are the characteristics of this profession? (Chapter 1)
- How will I develop the skills and competencies? (Chapter 2)
- What support networks exist? (Chapter 3)
- What competencies will I use? What activities will I undertake? What artifacts will I create? (Chapter 4)
- In what environments will I work? (Chapter 5)
- What is the outlook for this profession into the future? (Chapter 6)

This book gives you **research-based insight** into the technical communication profession and its practice, from education and careers, through workplaces and day-to-day activities, to possible future developments and prospects. It includes practitioner voices and has a dual focus on profession and practice. The excerpts and vignettes from the practice narratives suggest a range of entry routes and possible careers, sectors, workplaces, communities, processes, activities, and futures in this profession. They also help you to understand what technical communicators do in the modern, international economy. The final three chapters describe theories and research methods that informed the book and explain what the content implies for the strategic direction of the profession.

Who Should Read this Book?

This book is specifically targeted toward undergraduate and graduate students studying technical communication, early-career practitioners, and individuals considering a career change to technical communication. The content will also be relevant for educators in this or a related field and academic or industry researchers of the profession and its practice.

Undergraduate and Graduate Students

The book will help you to explore education and training opportunities and to understand how and where you can practice technical communication. Vignettes will enable you to explore education and training opportunities and to have a more authentic understanding of practice. The book should also enable you to consider which aspects of practice most appeal to you and which you might pursue in your career. The section on theory and research enables you to recognize and reflect on the links between research, theory, education, and the profession and the practice of technical communication.

Early-Career Industry Practitioners

If you are an early-career technical communicator, the book will help you to establish your career. If you have worked in only one role, one sector, or one region, it will provide a broader, international perspective on the scope of the profession through discussion of education, careers, workplaces, and practice. The vignettes may be relatable to your current practice and may also enable you to broaden your career vision. The discussion of strategic directions for the profession will help you to recognize professionalizing activities you can undertake.

Lateral Entrants to Technical Communication

If you work in another discipline and are considering a move into technical communication, this book will help you to make the transition. Its detailed information about routes to practice, relevant communities, and key activities can inform your decision-making. The vignettes will help you to visualize your future involvement in the field.

Educators

If you are a technical communication educator, several chapters of the book can help you in your teaching and your course and program development. Chapter 2 discusses international educational offerings and projects. It will help you to consider how to plan and design courses and programs. Through examples in several chapters, the book explores different types of industry and global locations of work. This content will help you to prepare your students for work within these different environments. The book is suitable as a primary text for courses about the profession and practice of technical communication because it includes information about a broad range of activities with an international outlook. The explanations of how I gathered and analyzed data will be relevant secondary texts for courses on research methods.

Academic and Industry Researchers

If your work primarily involves research, Part III will be particularly relevant to you. Chapter 7 discusses relevant theories and studies about professions and practice, including technical communication studies. Chapter 8 explains the research questions the book set out to examine and the methodological and analytical approaches. Chapter 9 summarizes insights and research and strategic directions. If your research involves professional or practice-based studies, you may be able to explore the proposed research directions.

Summary of Chapter 1

- The focus of this book is the profession and practice of technical communication in industry and internationally.

- Technical communication is a varied and diverse profession, but it has features that make it distinctive and recognizable. These features include a user-centered focus on developing content that enables people to interact with information, devices, or services.
- The profession's historical roots date back to the early twentieth century. Practices and technologies have evolved considerably in the past few decades, and they will likely continue to do so.
- Technical communication research and theory are interdisciplinary. Some research directions have had a considerable impact on practice.
- Several studies have identified a gap between research and practice, however.
- Multiple job titles are linked to technical communication, and our profession has connections with many related fields, like localization and translation, instructional design, and user experience (UX).
- Whether you are a student of technical communication, an early-career practitioner in this or a related field, a teacher, or a researcher, you will find this book useful for your current and future work.

Discussion Questions

1. This chapter includes several professional and academic definitions of technical communication. Find definitions from other sources (e.g., blogs or textbooks). Which is your preferred definition and for what reasons?
2. Using your preferred job search engine, conduct searches for "technical writer," "technical communicator," "information developer," "UX writer," and any other job title you are interested in. Consider these questions:
   How many jobs are advertised for each role? In which regions?
   What skills and competencies are required for the roles advertised?
   What industrial sectors, tools, and applications are mentioned?
   What additional information do the advertisements provide about the roles?
3. Review the Table of Contents and abstracts in the most recent issue of any technical communication journal you can access. Based on this review, identify the theoretical and research directions of the articles in that issue.

References

1 St.Amant, K. (2015). Introduction: Rethinking the nature of academy-industry partnerships and relationships. In T. Bridgeford & K. St.Amant (Eds.), *Academy-industry relationships and partnerships: Perspectives for technical communicators* (pp. 11–18). Routledge.
2 Rude, C. D. (2009). Mapping the research questions in technical communication. *Journal of Business and Technical Communication, 23*(2), 174–215. https://doi.org/10.1177/1050651908329562
3 Hayhoe, G. F., & Conklin, J. (2011). Conclusion. In J. Conklin and G. F. Hayhoe (Eds.) *Qualitative research in technical communication* (pp. 376–383). Routledge.

4 Henry, J. (2000). *Writing workplace cultures: An archaeology of professional writing*. Southern Illinois University Press.

5 Wenger, E. (1998). *Communities of practice: Learning, meaning, and identity*. Cambridge University Press.

6 Society for Technical Communication. (2020). *Defining technical communication*. https://www.stc.org/about-stc/defining-technical-communication/

7 Tekom Europe. (2020). *Making sense of complex matters*. https://www.technical-communication.org/technical-communication/defining-technical-communication

8 Australian Society for Technical Communication. (2020). *What is technical communication?* https://www.astc.org.au/about-technical-communications

9 Technical Communication Body of Knowledge. (2019). *Definition of technical communication*. https://www.tcbok.org/technical-communication-body-of-knowledge/definition-of-technical-communication/

10 Schriver, K. (2012). What we know about expertise in professional communication. In V. W. Berninger (Ed.), *Past, present, and future contributions of cognitive writing research to cognitive psychology* (pp. 275–312). Psychology Press.

11 Kynell, T. (1999). Technical communication from 1850 to 1950: Where have we been? *Technical Communication Quarterly*, *8*(2), 143–151. https://doi.org/10.1080/10572259909364655

12 Connors, R. J. (1982). The rise of technical writing instruction in America. *Journal of Technical Writing and Communication*, *12*(4), 329–352.

13 Staples, K. (1999). Technical communication from 1950 to 1998. Where are we now? *Technical Communication Quarterly*, *8*(2), 153–164. https://doi.org/10.1080/10572259909364656

14 O Hara, F. M. (2001, May). A brief history of technical communication. *Proceedings of the Annual Conference of the Society for Technical Communication*, *48*, 500–504.

15 Kirkman, J. (1996). From chore to profession: how technical communication in the United Kingdom has changed over the past twenty-five years. *Journal of Technical Writing and Communication*, *26*(2), 147–154. https://doi.org/10.2190/6P5V-TFAC-0XP7-FNPH

16 Carroll, J. M. (1998). Reconstructing minimalism. In J. M. Carroll (Ed.), *Minimalism beyond the Nurnberg funnel* (pp. 1–18). MIT Press.

17 Naughton, J. (2016). The evolution of the Internet: from military experiment to General Purpose Technology. *Journal of Cyber Policy*, *1*(1), 5–28. https://doi.org/10.1080/23738871.2016.1157619

18 Mirel, B., & Spilka, R. (2002). Introduction. In B. Mirel & R. Spilka (Eds.), *Reshaping technical communication* (pp. 1–6). Lawrence Erlbaum Associates.

19 Hughes, M. A., & Hayhoe, G. F. (2008). *A research primer for technical communication: Methods, exemplars, and analyses*. Lawrence Erlbaum Associates.

20 Abend, G. (2008). The meaning of 'theory.' *Sociological Theory*, *26*(2), 173–199. https://doi.org/10.1111/j.1467-9558.2008.00324.x

21 Rude, C. D. (2015). Building identity and community through research. *Journal of Technical Writing and Communication*, *45*(4), 366–380. https://doi.org/10.1177/0047281615585753

22 Schriver, K. (1997). *Dynamics in document design: Creating texts for readers*. Wiley.

23 St. Amant, K., & Melonçon, L. (2016). Reflections on research: Examining practitioner perspectives on the state of research in technical communication. *Technical Communication*, *63*(4), 346–364.

24 Boettger, R. K., & Friess, E. (2016). Academics are from Mars, practitioners are from Venus: Analyzing content alignment within technical communication forums. *Technical Communication*, *63*(4), 314–327.

25 Society for Technical Communication. (2010). *"Technical writer" officially a distinct profession per US government*. https://www.stc.org/notebook/2010/01/28/technical-writer-officially-a-distinct-profession-per-us-government/

26 Shalamova, N., Rice-Bailey, T., & Wikoff, K. (2018). Evolving skillsets and job pathways of technical communicators. *Communication Design Quarterly, 6*(3), 14–24. https://doi.org/10.1145/3309578.3309580

27 Anschuetz, L., & Rosenbaum, S. (2002). Expanding roles for technical communicators. In B. Mirel & R. Spilka (Eds.), *Reshaping technical communication* (pp. 149–164). Lawrence Erlbaum Associates.

28 John Benjamins Publishing Company. (n.d.). *Information design journal.* https://benjamins.com/catalog/idj

29 Halverson, K. (2008, December 16). The discipline of content strategy. *A List Apart.* https://alistapart.com/article/thedisciplineofcontentstrategy/

30 Hassenzahl, M., & Tractinsky, N. (2006). User experience—A research agenda. *Behaviour and Information Technology, 25*(2), 91–97. https://doi.org/10.1080/01449290500330331

31 Hassenzahl, M. (2008, September). User experience (UX) towards an experiential perspective on product quality. In *Proceedings of the 20th Conference on l'Interaction Homme-Machine* (pp. 11–15). ACM. https://doi.org/10.1145/1512714.1512717

32 Lebson, C. (2016). *The UX careers handbook.* CRC Press.

33 Minacori, P., & Veisblat, L. (2010). Translation and technical communication: Chicken or egg? *Meta: Translator's Journal, 55*(4), 752–768. https://doi.org/10.7202/045689ar

34 Lattimore, D., Baskin, O., Heiman, S. T., & Toth, E. L. (2012). *Public relations: The profession and practice (*4th ed.*).* McGraw Hill.

35 Rosselot-Merritt, J. (2020). Fertile grounds: What interview of working professionals can tell us about perceptions of technical communication and the viability of technical communication as a field. *Technical Communication, 67*(1), 38–62.

2 Technical Communication Education and Professional Development

Introduction

Although technical communication has existed for as long as people have needed instructions, it did not emerge as an academic discipline until the mid-twentieth century. Because it is a new profession, there is no one educational path to becoming a technical communicator. You can qualify specifically for this profession. If you already have a degree majoring in another discipline, you can move into technical communication. There are multiple training and professional development (PD) possibilities. This chapter will give you insight into some of the international education and training opportunities in this profession.

This chapter starts with a **time line** of how technical communication education developed. The timeline shows the key events that led to educational developments, and it helps you to understand the educational context.

Most of the chapter focuses on **formal education**, **professional development**, and **certification**, including international offerings and projects. Although the boundaries between technical communication and other disciplines can be fuzzy (as discussed in Chapter 1), the focus of this chapter is education and training that is grounded in technical and professional communication. I distinguish between formal academic education through the higher-education system and training and professional development, such as short-term courses and learning on the job.

You will find examples from the **practice narratives**, and from projects, a program spotlight, online sources, and research. These examples offer insights into current education and training routes, and they show how technical communicators learn in education and work settings. A **vignette**, a story based on the practice narratives, describes an early-career technical communicator's experiences of education and her first steps in her career.

After reading the chapter, you will be able to identify the range of education and training opportunities offered internationally.

- If you are a **technical communication student**, you will be able to compare your program, courses, and educational experiences with the examples. You will also be able to identify resources that can strengthen your understanding of training and education in this profession.
- If you **work in industry as a technical communicator**, you will be able to relate your experiences to the education and training opportunities and projects, and you will be able to plan your future education and training goals.
- If you are **considering a career change into technical communication**, you will be able to find education and training programs and resources that could facilitate the transition.

Development of Technical Communication Education

Although specialized writing courses were included in engineering programs from the early 1900s, Rensselaer Polytechnic Institute offered the first academic program in technical communication in the 1950s.[1] The time line in Table 2.1 shows key developments in technical communication education in the past 150 years. As you can see from the time line,

- This academic discipline has evolved continually since it emerged in the mid-twentieth century.
- Fundamental changes in technology and society influenced the development of education and training. These changes included the widespread adoption of personal computers (PCs) in the 1980s and the development of the internet in the 1990s.
- Until the 1990s, the academic field was most prominent in the US. New educational offerings and projects have started to develop worldwide since then.

Practicing Without Academic Qualifications

Because technical communication did not have a strong academic identity until the late twentieth century, many individuals throughout the world have moved into the profession from other specializations and without studying for technical communication qualifications. For decades, many technical communicators (then more commonly known as technical writers) practiced without having academic qualifications in this discipline and without undertaking formal specialized training. This was often, and continues to be, a second career.

An early technical communication discussion forum, TechWhirl (see Chapter 3 for more information about this forum), regularly featured heated debates

Table 2.1 Time line of educational developments in technical communication

Date	Development
Late 19th century and early 20th century	Changes in engineering education in the late 19th century led to Humanities subjects being sidelined to make space for a more technical curriculum. The perception that engineers had serious problems with written communication led to a demand for specialized writing courses for engineers.[2]
Post–World War II	Technical writing became an important skill:
	"For each new airplane, gun, bomb, and machine needed a manual written for it, and the centrality of the lucid explicator was obvious as never before" (p. 341).[2]
	As a result, academic programs were needed to train technical communicators.
1950s and 1960s	Professional associations were founded, and Rensselaer Polytechnic Institute developed the first master's program in technical and scientific writing.[1]
	Two journals began to publish technical communication scholarship: *Technical Communication* and the *IEEE Transactions on Professional Communication*.
1970s	The profession of technical communication and the professionalism of its instructors became important, leading to a move away from service teaching. *A humanistic rationale for technical writing*[3] argued that technical communication needed to establish its own academic identity within the Humanities.
	New journals appeared: the *Journal of Technical Writing and Communication* and the *Technical Writing Teacher* (this journal was relaunched as *Technical Communication Quarterly* in 1992).
	The Council for Programs in Technical and Scientific Communication (CPTSC) was established. This organization supports technical communication program directors.
1980s	The boom in the personal computer industry led to an increase in the number of technical communication programs; the number of programs did not meet the demand for writers created by thriving desktop computer sales, however:
	"many unqualified people declared themselves to be technical writers by virtue of backgrounds in journalism or English that did not necessarily give them the specialized knowledge required to produce usable documentation" (p. 212).[4]
	The *Journal of Business and Technical Communication* was established.
1990s	The rapid pace of development of the internet and its related technologies, together with the impact of globalization, created new challenges and opportunities in technical communication education. Curricula became more diverse to help prepare students for new, digitally enabled content development roles.
	Full-time specialized technical communication programs developed in Europe in the 1990s,
	"at varying levels … in England, Scotland, Ireland, Denmark, France, Germany, the Netherlands, and Sweden" (p. 150).[5]
	Programs outside the US were developing "rapidly world-wide but unevenly" (p. 111).[6]

(Continued)

Table 2.1 (Continued)

Date	Development
2000s	Partially or fully online programs in technical communication began to emerge, as internet technologies and infrastructure became more advanced.
	More programs developed internationally. An edited collection about teaching intercultural rhetoric and technical communication included chapters on technical communication education in France, India, Ireland, Israel, and New Zealand.[7]
	Programmatic Perspectives, the journal of the CPTSC, first appeared.
2010s	The Society for Technical Communication and tekom developed and refined separate certification programs to certify skills of practitioners without academic qualifications in technical communication.[8]
	Projects such as the Technical Communication Body of Knowledge and TecCOMFrame (both discussed later in the chapter) were established to support education.
	The Association for Computing Machinery Special Interest Group on Design of Communication (SIGDOC) launched a new journal, *Communication Design Quarterly Review*.

about whether technical communication education was necessary. Seasoned professionals who had trained in other disciplines often argued that a successful career did not have to begin with, or ever involve, a technical communication degree. Examples from the practice narratives show how such careers began and later flourished. As you saw in the time line in Table 2.1, the adoption of PCs was an important stage in the evolution of technical communication education, as academic programs emerged to fill a new labor market gap. In the absence of these programs in many regions, individuals like this respondent moved into the profession without undertaking a related academic program:

> I got into Technical Writing almost 30 years ago by accident. I started working at a company as a procedures writer just as PCs started being incorporated into regular businesses. What had formerly been corporate procedures, etc. became application user manuals and training documentation used by our trainers.

Another respondent described the common situation of taking training courses rather than formal education:

> As to education that led to my current role, there's no education like being thrown into the fire and told to swim. Ha. I'm intentionally mixing metaphors. Basically, I had no formal training except for tools training now and then, project management training, working directly with SMEs [subject-matter experts], and just honing my skills by doing the job.

Many members of the profession still lack certification or academic qualifications, but there is increasing recognition that specialized education and training are important.

The Value of Education for the Profession

Technical communication researchers have explored how a lack of academic qualifications can be a disadvantage for practitioners and for the profession as a whole. If unqualified people, however skilled they are, work in technical communication roles, this work may not be recognized as having unique or complex skills. It is less likely to be well paid, and the job can change from one role to another. The quality of work may also be uneven.

Nowadays, more academic and training programs are available in many countries. If you are new to this field of work and you do not have a formal qualification or certification, you may find it difficult to evidence your competence, skills, or professional expertise. Because degrees are offered in many countries worldwide, you will likely be competing in the labor market with applicants who have degrees in technical communication.

For many reasons, academic qualifications are important in any profession. Research about established and new professions shows that academic programs:

- Help to define the skill set.
- Help to ensure that only individuals who have a certified level of competence can practice in the profession.
- Confer authority and prestige on the professional and the profession.
- Distinguish specialized from non-specialized work.

Through all these benefits, education and training can increase our professional identity and enable us to recognize our value. In the workplace, access to training and PD increases our sense of being valued by our employer. One practice narrative respondent complained about a lack of investment in their PD. The language in this excerpt suggests that they feel undervalued. You'll see that they used the term "value" twice:

> While my company will pay for my STC [Society for Technical Communication] membership, they did not see the value in sending me to the STC Summit, until this year, but there were extenuating circumstances that drove that decision I think. I would like to go to teacher training that we provide for my own PD, but I've not been able to convince anyone that that's a good value either. PD is a sore spot for me.

In Chapter 7, a longer discussion about professionalization theory explains why education is important to a profession's status and prestige, to a professional's identity, and to the professionalization aspirations of many new and emerging professions.

Savage argued that, as long as there is no requirement for qualifications and certification in technical communication, some practitioners have only "basic aptitudes or backgrounds in technical subject matter." He compared this situation with "allowing people to perform surgery who are 'good with their hands'" (p. 2).[9]

For these reasons, education and training in technical communication have been priorities for the STC for more than two decades. Tekom Europe, the European Association for Technical Communication, is also working to develop and increase the number of academic programs throughout Europe.

International Education and Training Programs

Technical communication programs that cater for different types of students are now available in many regions of the world and at several academic levels: from certificate, diploma, and bachelor's and master's levels to advanced research degrees like the PhD. Service courses are also common. These service courses are usually part of a degree program in another discipline (like engineering, languages, or media). They cover basic concepts and skills. Table 2.2 shows a snapshot of the range and types of programs and training options available internationally. Because programs change all the time, the table does not list individual programs and courses.

As Table 2.2 shows, program offerings differ throughout the world:

- Many educational opportunities exist at all levels in technical communication in North America, particularly in the US.
- European programs have begun to develop in the past two decades. In that region, the academic sector is strongest in Germany. In several European countries, including the UK, Italy, Poland, and Spain, no full higher-education programs are available.
- There are no recognized programs in South America or Africa.
- Although some relevant courses are offered in universities in Australia (e.g., at the University of South Australia), full programs are not available. One New Zealand institution offers a Graduate Certificate and Diploma in Information Design, and another offers a course in Professional and Technical Writing.
- In Asia, programs are offered in a small number of Chinese universities, some with ties to Western universities.

Finding and Choosing a Study Program

In the US, technical communication programs are based in different university departments, including English, Writing/Composition/Rhetoric, Engineering, Media Studies, and Communication Studies.[15] In Europe, programs are more likely to be in Language, Communications, or Media departments. Several websites, like Mastersportal.com (see Figure 2.1), enable you to search for programs worldwide.

Program Titles

Many job descriptions now begin with the word "content," like "content strategy," "content engineering," "content development," and "content management" (see Chapter 5). Worldwide, program titles are beginning to change to

Table 2.2 International programs and professional development in technical communication

Region	University programs	International professional development courses
North America	In the US, full programs are available at all levels, from two-year college to degree major, and certificate through diploma and master's to PhD. Programs are offered in multiple delivery modes (online, campus-based, hybrid, distance). Melonçon and Henschel identified 65 undergraduate programs with majors in technical and professional communication offered in the US, and a total of 185 undergraduate programs that included a technical communication component.[10] Melonçon identified 84 postgraduate technical communication programs on offer in the US.[11] In Canada, programs are offered at certificate, bachelor's, and master's levels in several institutions, including Mount Royal University and Concordia University.	Two formal certification programs exist: • The Society for Technical Communication's (STC's) Certified Professional Technical Communicator (CPTC) program. • Tekom Europe's certification program. Relevant MOOCs (massive open online courses) are offered by several providers, including MIT, Coursera, and Udemy. In the UK, Cherryleaf offers online and classroom-based training courses, including courses in intermediate and advanced technical communication, tools, and writing skills. In Australia, Engineering Education Australia offers a course in writing documentation that contributes to chartered status, ATTAR (Advanced Technology Training and Research) offers two-day public courses in technical writing, and Australian Online Courses offers a Certificate in Technical Writing. In Israel, training courses are run by private companies including CowTC and Our Best Words. Web-based opportunities for training, are available from, among other providers, • The STC. • LinkedIn Learning. (See Appendix A for details of all these organizations.)
South America	No full academic programs are offered.	
Australia and New Zealand	The ARA Institute of Canterbury in Christchurch, New Zealand offers a Graduate Certificate and Graduate Diploma in Information Design, and Open Polytechnic, New Zealand offers a course in Professional and Technical Writing.[12] Some editing, publishing and communication programs are offered in Australia, but none specifically in technical/professional communication	
Asia	In China, courses are offered at a small number of institutions, including Peking University, Nankai University, and Xidian University. Some of these programs have ties with Western universities.[13] In India, Pandit described a Post Graduate Diploma in Technical Communication (PGDTC) offered by Pune University.[14] This program appears to have been discontinued.	
Europe	In Germany, several institutions offer master's degrees, and a full undergraduate degree program is offered by the Karlsruhe University of Applied Sciences (Fachhochschule). (See the "Program Spotlight" later in this chapter.) In France, an undergraduate major in technical communication is offered at the Université de Limoges, and master's degrees are offered at institutions, including Université de Paris, Université de Bretagne Occidentale, Université de Strasbourg, and Université Rennes 2. Postgraduate programs are also offered in Austria, Finland, Ireland, Sweden, Switzerland, and the Netherlands.	
Africa	No full academic programs are offered.	

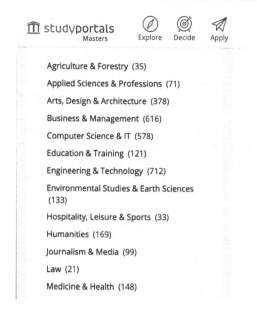

Figure 2.1 Disciplinary results for "technical communication" at https://www.masters-portal.com (accessed November 13, 2020)

match job titles in industry. It may be time for a "rethinking of what we call our programs and degrees" (p. 24)[16] because job titles are changing. Some programs include "technical communication" and another term in the title. For example, the master's program at the University of Limerick, where I work, is called "Technical Communication and E-Learning" to emphasize that students learn about both technical communication and instructional design/e-learning. Programs that previously included "technical communication" in the title may now have other labels like "information design," "content strategy," and "digital content" to reflect moves toward digitalization, user-centered design, and content development. These are still technical communication programs.

Examples of alternative program titles:
  Communication and Media Management (University of Applied Sciences, Karlsruhe)
  Information Design (Mälardalen University, Sweden).

Academic Degrees: Certificate, Master's, and PhD

The TecCOMFrame project (described later in this chapter) identified programs at six different levels, from a full bachelor's degree that provides broad and deep coverage of multiple technical communication subjects to a service (or specialization) course or subject stream within another study program (such as a languages degree). (If you want to learn more about the six levels, see Karreman et al.[17])

A recent special issue of the STC's *Intercom* magazine included articles outlining the benefits of postgraduate study at three graduate levels: certificate, master's, and PhD. These levels are common throughout the world. In summary,

- A **graduate certificate** in Technical Communication includes about five courses, takes a year to 18 months to complete, and tends to focus on workplace skills. This type of program might be a good fit if you have trained in another discipline (like engineering or English) but you find yourself in a role that involves professional or technical communication.[18]
- A **master's program** in Technical Communication is a good choice if you want to earn a higher degree, develop critical thinking skills, and get a deep understanding of subjects relevant to technical communication. In many cases, these programs are open to applicants with a bachelor's degree in any discipline. You need to be in a position to invest time in postgraduate work (typically up to ten courses, taken over the course of one to three years depending on the institution and the study mode).[19]
- A **PhD** in Technical Communication is an indicator of research expertise in an area of the profession and is an internationally recognized prestige marker.[20] Doctoral work can take several years to complete. Doctoral students must commit to researching a niche area in depth. In the past, a PhD typically led to an academic career, but PhDs in industry are increasingly common.

If you want to change careers or return to the workplace after a career break, a master's degree in technical communication enables you to return to education to qualify specifically for this field of work. In our master's program in Limerick, as in many similar programs, some students previously trained as teachers, journalists, engineers, and in many other disciplines. Other students come to our program after a career break (usually to raise children, but some students have other caring responsibilities). A practice narrative respondent described a similar path:

> While on a career break raising my children I taught music and drama before upskilling by completing an MA [master of arts] in Technical Communication in order to re-enter the workforce on a full-time basis.

Programs at different levels have different potential benefits. You should choose a program based on, among many other factors, the following:

- Your academic interests.
- The stage you are at in your education.
- The labor market in your region.

- Whether you want to study online or through a campus-based program.
- The specialization you are interested in pursuing.

Choosing a Program: A Vignette

June is an early-career technical communicator, and here she describes her experiences of education, initiation into practice, and initial training and PD experiences. "June" is a composite character, and her story is based on data from the practice narratives. Like respondents to the practice narratives survey, she reflected on her strengths, her interests, her learning preferences, and her education level and selected a program that matched her needs.

My favorite subject at school was English. We had a career assessment at school, and I scored highest in abstract thinking, but I wasn't sure how to apply this knowledge to selecting a study program. I took a double major in English Literature and Media Studies for my undergraduate degree. At some point during that program, I discovered that I was good at explaining complex concepts, especially through writing. During my last year at university, I did an apprenticeship in a large multinational software company and that company hired me after graduation. Even though I hadn't trained in it and it was not a profession I was familiar with, the job I was hired to do was technical writing. My main work, initially, was to rewrite content written by engineers, using MS Word. Gradually, I also learned to use the authoring tools and the content management system that the full-time technical writers used and to write other types of content.

At first, everything seemed complicated, but the job started to make sense after a few months. I received some formal training, especially on company policies. I also had access to webinars and corporate e-learning programs. I shadowed one colleague for a couple of weeks, and afterwards she was assigned to be my mentor. This mentoring involved weekly meetings where she answered my questions and gave me feedback on my writing specifically but also on my work performance generally. She was open to me contacting her between meetings, but I didn't feel confident doing so. I tried to rely on my own resources to learn. I learned a lot from listening to and observing colleagues on my team, especially about how to participate in a meeting and how to interview subject-matter experts. The corporate style guide really developed my understanding of key technical writing concepts like consistency, correctness, and accuracy. It also enabled me to make writing decisions where many options were possible. Because I had studied English Literature, I was interested in writing, but this way of writing with such precision and conciseness was new to me. As well as learning about how to do the job of technical writing, I also learned a lot about working in a corporate environment: being effective in meetings, managing and tracking my time, juggling projects, and collaborating in a team.

I worked in that organization for a year following the internship. I left because I knew the job well and was ready to move on to another technical communication role. I was interested in writing content for users of technology and getting to know more about user experience (UX), instead of rewriting developer content. When I started to look for work in UX writing, I discovered that almost all job advertisements required a technical communication, UX, human-computer interaction, information design, or similar academic qualification. I started to research graduate study options and found quite a few technical communication programs, including some online options that would have enabled me to keep working while I studied. Instead, I decided to study full-time and on-campus rather than taking an online program. I was able to afford the fees, and I wanted to immerse myself in learning for a year. The master's program I chose was in technical communication, with some specialization courses. There were just 15 students on the program the year I took it, so we really got to know one other and our instructors. The core courses were offered at night, so I was able to keep a part-time job. I chose this program because it was in my hometown, and it offered an elective course on human factors and usability, which was the specialization where I wanted to work. The program also covered writing and rhetoric fundamentals, theory, research, and interaction design. There was no internship, but the univer-sity had a lot of industry ties, employers came to campus, and we were all recruited directly from the program.

I graduated two years ago and began working as a Junior UX Writer with a large, well-known company. I had a week of initial training and, again, a mentor was assigned to me. This time, he was more involved in my daily work.

The types of academic qualification that are relevant to my role are varied because even some that are not obviously linked could be helpful, and people on my team have varied educational backgrounds that make it interesting and exciting for us to work together. I wouldn't have gotten this job without my MS degree, but the degree is not enough. I have to keep updating my skills. I try to participate in projects that will help me to develop different skills and to gain expertise in different domains, like writing, tools, design, UX, and leadership. I'm in contact with many of my former classmates from the MS program, and we have our own social media groups. It's a great network. I've also recently joined a professional association that I hope will be valuable for keeping my skills current. For future training and education, I may consider another formal program, but for now I'm doing whatever relevant training that I can fit in with work and that my employer will fund.

June's reflection and decisions about education and training might help you to think about how to choose a study program or plan your professional development.

Typical Program Content

Although the program June described in the vignette covered content that is common, no standard technical communication curriculum exists at any academic level. A survey of 142 programs in the US found no "representative" curriculum.[21] In regulated professions (e.g., law and many health professions), the academic curriculum must include specific courses for accreditation. A graduate who has not completed an accredited program cannot practice in the profession. That is not the case in technical communication. There are some common courses, however.

- At the **undergraduate level**, core courses often include introductory and advanced technical communication, information design, editing, genre, and web development. Courses also emphasize digitalization and social media, information design, and internships.[10]
- At the **graduate level**, one study analyzed 84 technical and professional communication programs in the US. Commonalities included an "Introduction to the Field" course, a research methods course, and an internship. This study found postgraduate programs to be more diverse than undergraduate programs.[11]

Because of continual changes in technology and in organizational structures, the skill set is changing continually, and programs have to keep up with those changes.[22] Recent studies about technical communication programs and curricula point toward several trends:

- Technical communication programs now often include courses on web and applications development, interaction or UX design, intercultural communication, accessibility, and content management.
- More programs also include courses on structured writing and topic-based writing, reflecting the widespread adoption of this type of writing in industry.
- In Europe, programs regularly include content on translation, intercultural communication, and international communication.
- Ethics and how to write and design ethical content have been a focus of technical communication curricula for decades (see, for example, Kienzler[23] and Markel[24]).
- Social justice is a more recent concern, especially in North America. Social justice content aims to help "technical communicators understand how to avoid sexism, racism, able-ism, age-ism, and other prejudices within research, teaching, and professional practices" (p. 5)[25].

Another trend is toward more online programs and courses. The availability of partially or fully online programs, especially at the graduate level, has increased educational access. An online program might suit you if you are working full-time or if you cannot access a campus-based program.

Examples of Education Paths

In the practice narratives, participants wrote about their education and early work experiences (see Appendix B for the survey questions). Sixty-two individuals based in 13 countries submitted narratives describing their educational backgrounds, initiation into technical communication roles, and ongoing engagement with training and PD. Because respondents provided some basic demographic data (age ranges, country base, and gender), I was able to ascertain some emergent patterns in educational paths, but given the small number of respondents, these trends do not represent all technical communicators. They do, however, suggest possible directions for education and for future research.

- All the respondents were highly educated regardless of the country they lived in, and most had a university education to at least the bachelor's level. However, the bachelor's specialization was rarely in technical communication. The range of undergraduate specializations was vast; respondents described educational backgrounds in, among other disciplines, languages, computer science, business, social science, and biomedical science.
- Three respondents (one in Sweden and two in the UK) did not have an undergraduate degree and a fourth (based in the US) had "some college" but did not state the level. All four were in their fifties. Otherwise, all respondents had at least a bachelor's degree.
- Just two respondents had an undergraduate degree in technical communication, but more than half (33 respondents) had a postgraduate qualification in technical communication, most commonly at the master's level (29 respondents). Five individuals had more than one master's degree, one had completed a doctorate, and another respondent was planning to undertake a doctorate in the future. Although this level of postgraduate education is unlikely to be representative, it suggests a trend of more programs and more individuals qualifying specifically to work in this profession.
- Respondents under the age of 40 were much more likely than older respondents to have completed an academic program, usually at the postgraduate level, in technical communication.
- Of the 23 respondents based in countries where English is not the first language, 11 had completed a language-related undergraduate degree (e.g., foreign languages, translation, or linguistics). By contrast, five of the 37 respondents based in English-speaking countries had a primary degree in foreign languages. This pattern seems to suggest the dominance of English-language technical content and points to the links between language work and technical communication in education and industry in many European countries.
- Twelve respondents were currently undertaking academic study or had plans for study in the near future. Most of the study disciplines were in

related areas such as software design, online learning, innovation management, and UX. A couple of respondents were doing, or planning to do, unrelated degrees (in fine art and theology).

Program Spotlight: Communication and Media Management at the University of Applied Sciences, Karlsruhe, Germany

Sissi Closs is Professor of Information and Media Technology, former Director of the master of science (MSc) in Communication and Media Management, Equal Opportunities Officer, and International Representative at the Karlsruhe University of Applied Sciences in Karlsruhe, Germany. Sissi developed and led technical communication programs at the Karlsruhe University of Applied Sciences since they were introduced in 1997. I interviewed her about how the programs changed during that time.

How did you first get involved with technical communication education at Karlsruhe?

I come from a Computer Science background. In 1987, I founded the first software documentation company in Germany. I could see the high potential for people with interdisciplinary skills (writing, programming, and design). I had to teach all our employees from scratch in this new discipline. Thus, I learned intensively what is needed and how it can be taught efficiently to people from any discipline.

I set up the program at Karlsruhe with two colleagues who were language specialists (in German Studies and Translation). Our program in Karlsruhe, which started in March 1997, was the second technical communication study program in Germany (the first one was in Hannover). Initially, we offered an eight-semester "Diplom" program. This was a qualification comparable to a master's degree. The program was called Technische Redaktion (Technical Writing). We had eight students in the first year. We taught four core subjects: language, information and media technology, visualization/design, and basic engineering technology. These four subjects are still at the core of our programs, but a lot has changed since then.

We ran a second program for people already working in industry, and who already had a degree. This was a short, four-semester, postgraduate program. Students had a wide variety of backgrounds, which corresponds to the interdisciplinary character of our field. The students appreciated the postgraduate program very much. Many got much better jobs than they had been able to achieve with their previous qualification.

How have the programs changed in the past two decades?

In the early days, students took technical courses in the Engineering Department with students from engineering programs (who were training to be engineers). This model did not work. Our students didn't feel at home, and the engineering students didn't accept our students because they were not training to be engineers. This led to our first big change: we created separate technical courses for technical writing students, to ensure they were at the right level and that the students learned what they needed to learn. Demand for technical writers in industry was strong at the time, and the student numbers were good, but the study program was not attracting young people. "Technical writing" seemed old-fashioned, and wasn't well known in the public domain. We had enough students but we wanted to have more applications than places, and to be able to choose the most suitable applicants.

This led to the next big program change in the Winter semester of 2005. We reorganized our programs to offer degrees at bachelor's and master's levels. [This development was in line with changes that were happening throughout the higher education system in Europe.] We also renamed our programs. The new title, for both the BSc and MSc, is Communication and Media Management (Kommunikation und Medienmanagement). This title reflects the emphasis on communication, and on managing media, especially digital media. Some applicants think this is a management program, but it's not. The first part of the title is about communication, and the second part is about managing media. The rebranding has been very successful. Between 1,200 and 1,500 people apply for the bachelor's program annually, and we accept up to 90 new students annually. The study program has become well known in Germany. The MSc is also very popular. We usually have 100–120 applicants, and we accept 20 new students annually.

What is the structure of the bachelor's program?

The curriculum is still based on the four pillars that we started with in 1997: language, information technology, media, and visualization, and it is a seven-semester program. We still offer the program once a year, starting in the Winter semester. Of course, a lot has changed. We review the program content regularly, and we are currently running the fourth version of the curriculum. (You can see the module titles in Table 2.3.)

We have no elective modules at the moment, even though many students would like to be able to specialize. We can't offer electives because there is no space on the program. We have to make sure students study the core subjects. We would also need more teachers to be able to offer electives.

Table 2.3 Structure of the bachelor of arts (BA) in Communication and Media Management

Semester 1	Semester 2
Information Technology (IT) Basics A	IT Basics B
Linguistics and Language Skills	Text Linguistics A
Design	Language and Design A
Technology A	Technology B
Technical Documentation A	Technical Documentation B
	Internet Technologies

Semester 3	Semester 4
Information Technology	User Assistance
User Experience	Information Management
Technical Documentation C	Media Production
Technical Documentation D	Terminology Management
Language and Design B	Technology C
Text Linguistics B	

Semester 5	Semester 6
Practice Placement	Information Systems A
	Technology D
	Visual Communication
	Intercultural Communication
	Language Management

Semester 7
Information Systems B
Information Architecture
Media Management
Thesis Module
Final Seminar

What is the focus of the master's program?

The MSc in Communication and Media Management is a three-semester program. It is the only master's program in Germany for students who have a bachelor's degree in technical communication. But it's also open to students with degrees in other disciplines: for example, computer science, languages, and translation, as well as technical communication. Like in the bachelor's degree, we review the curriculum regularly. Our new curriculum will run from Winter 2020. (See Table 2.4 for a list of modules.)

Because only a few modules are obligatory, students can really specialize in one area. The variety of electives makes the program very heterogeneous for students.

Table 2.4 Master of science (MSc) in Communication and Media Management: Modules

Core modules	Elective modules
Media Law	Applied Linguistics and Language Management
Key Skills	Technology and Technology Didactics
Philosophy of Science and Epistemology	Visual Communication and Knowledge Transfer
	Semantic Information Management
Master's Thesis/Final Seminar	Technical Communication
	Media Engineering

Do students have opportunities for internships?

On the bachelor's program, students must undertake a six-month internship. That is a requirement in all University of Applied Sciences bachelor's programs. The internship can be in Germany or abroad, or students can do a project in another university instead. Master's students don't do an internship.

Do students write a thesis or do a final project?

In both the bachelor's and master's programs, students can do their final thesis with a company. Companies contact the university to propose thesis topics. We have a special university office that sets up these connections. Students can also propose to do a project in a company where they already work part-time, or we help them to find a project through our personal contacts. This is a good system because the students are involved in real projects that explore innovative research questions, and they get to use company resources. Sometimes, they even get paid to do the project.

What do graduates do after completing this program?

The range of options for our graduates is very broad. Many go into technical writing jobs. That is still the most common career. A lot of graduates also find work developing e-learning, training, learning materials, and other digital learning resources. Quite a lot of our people have founded their own documentation, or e-learning and training, or consulting companies. Others go into marketing. They sometimes work in marketing departments, e.g., for internships, where their competencies are recognized. These competencies include using language carefully, managing content, and managing digital channels like social media, web, and email. They are efficient organizers and develop professional materials, even though they are not marketing specialists because we have no marketing modules. Traditional marketing education does not teach these new competencies, so marketing departments are very impressed with technical communicators. Other graduates work in organizations where they help scientists to make content more understandable. They are the interface between scientists and laypeople.

What are your plans for the programs?

We want the program to remain successful, of course! But we are always reviewing what we do and making changes. One change we want to make now is to general-ize some module titles, especially in the higher semesters, so that they are flexible and we can teach new subjects: for example, deep learning, data analytics, and chatbots. These subjects didn't exist when we were developing the programs. It is important to be flexible. Some subjects are stable over time, e.g., grammar (even though we have had to change the German grammar content recently), or program-ming principles. But some content is new and changes all the time, like which programming languages to teach. By using more general module names we are free to bring in current subjects and to modernize the curriculum continually.

The spotlight shows how these programs have remained successful and competitive through flexibility, responsiveness, and adaptability.

Alternatives and Supplements to Formal Education

Although the practice narratives suggest a high level of engagement with tech-nical communication education among the respondents, we know from research and lore that for individuals in their forties and older, the path to becoming a technical communicator was often varied and meandering. People trained for and worked in other sectors, and they moved into technical communication "by accident." This is still sometimes the case, and there are diverse routes into the profession. You might be thinking about this as a second career if you trained in a related area or an entirely different area and you like writing and are enthusi-astic about technology. In that case, you are not alone.

Tekom Europe[26] estimated that in Germany just 20% of technical communicators have a university degree in the discipline, and Conklin and Hayhoe noted that the subscriber base of the *Technical Communication* journal "consists overwhelmingly of people who do not have degrees in technical communication and for whom the profession is a second or subsequent career" (p. 379).[27]

In many countries, no academic programs exist; therefore, most practitioners in those countries do not have technical communication qualifications. For example, if you live in the UK or India, your formal education was probably in another discipline and your technical communication training has been mostly outside of the university system.

Even if you have a degree in technical communication, you need to continue learning. In any profession, formal training in the higher-education system is only a starting point. Describing his first role in architecture, de Graff (p. 3)[28] explained how little his education had prepared him for this job:

I had graduated only six months earlier, and in many ways, that first job came as a shock, not so much because of the quality of the buildings that I worked on or the nature of the clients' decisions as because of the fact that practicing as an architect appeared to have nothing – and I mean *absolutely* nothing – to do with studying architecture. The first emotional state I recall having as a practicing architect was a feeling of utter uselessness. My technical knowledge fell far short of what was needed; at the same time, nobody was interested in the elevated philosophical considerations I had developed during my studies. I was both over- and under-qualified for my job.

This experience of feeling unprepared and simultaneously overqualified (usually in theoretical concepts) and underqualified (usually in practical skills) is common for new entrants to any profession and a reason why further PD is vital. All professionals must continue to develop their skills and keep up with new practices, tools, policies, standards, laws, and many other developments. The vignette showed how June had to keep her skills current even after completing a master's degree.

Because technical communicators must adapt to continual technology and practice developments, you will need to undertake training and professional development courses even if you have academic qualifications in technical communication.

Examples of Training and Professional Development

Various training and PD opportunities, ranging from occasional webinars to short courses taken online or onsite to longer-term courses, are available in most jobs. You can access PD courses through professional associations and private companies, in classroom-based or online modes, and for various levels of accreditation, including certification. Table 2.2 illustrates some of the many options for PD.

The practice narratives described many types of PD courses that respondents had taken, including in tools and technologies (e.g., in Darwin Information Typing Architecture [DITA], XML, FrameMaker, and MadCap Flare) and soft skills (e.g., in time management and presentation skills). Many respondents integrated this type of training into their work schedules, as in this example:

> I have completed short on-going training at least once a year while I've been working, and I intend to keep this up. This might be training in publishing software, communications skills, online short courses.

Whereas most narratives referred to training regimes organized and funded by an employer, three self-employed respondents described the challenges of directing, managing, and funding their training and PD. They reported motivation and making time for the training as being particular challenges. In these cases, the training might seem to be at the expense of completing paid work. However, if you are self-employed, as in the excerpt below, the longer-term view is that continual training enables you to take on more, and more varied types of, work in the future.

> I also want to learn more about API [application programming interface] documentation, but it's a struggle to find the time when I don't need this for my current projects. I just know it's coming up and I might be able to position myself for these [types of contracts] in the future.

As well as in-house training, the practice narratives included descriptions of several modes of informal learning:

- **Belonging to a professional association**. Most professional associations run training courses, both formal and informal, some paid and some free of charge. Typical types of training are podcasts, webinars, e-learning courses, and conferences. Professional associations are discussed in more detail in Chapter 3.
- **Reading books and journals**, including journals from professional associations. Respondents mentioned three types of books that they use for self-training: academic texts, popular science and psychology texts, and technical manuals.

- **Trying out tools**, applications, and programming languages (self-teaching).
- **Attending and presenting at conferences.** The biggest conferences in technical communication for industry practitioners are organized by the STC, tekom Europe, Institute of Scientific and Technical Communicators (ISTC), and the Center for Information-Development Management. For more information on these organizations and conferences, see Chapter 3.
- **Consulting mailing lists** and discussion forums. This respondent explained how they use forums to ask and answer questions:

  > I … use the ISTC forum when I have a specific question, and do my best to answer those posted by others.

- **Taking webinars and free on-line training** courses. Respondents noted specific online resources like Lynda.com and LinkedIn Learning as well as podcasts and videos.
- **Following and posting on social media.** Respondents mentioned LinkedIn as a tool for watching trends, and two narratives referred to following industry leaders and keeping abreast of trends on Twitter.

In a recent *Intercom* article, Carliner and Chen reported on how respondents to an STC census undertook similar types of PD.[29] Their list of PD activities also included **meetings as a training opportunity**.

Learning at Work

As well as learning from training courses, we all learn from observing and working with our colleagues. The communities-of-practice model[30] (discussed in more detail in Chapter 7) explains how people learn from each other in work settings. We learn how to practice through our interactions with others in real-world situations. For a technical communicator, for example, the difference between a university course and an on-the-job assignment is vast, and in the early stages of a new role, we constantly learn from others as we work. Lave and Wenger described how new hires learn a practice by observing how their colleagues work and interact. They called this type of learning "legitimate peripheral participation."[31]

Legitimate peripheral participation can be formal (e.g., through an internship or a workplace mentorship program) or informal, where we observe, reflect, and develop confidence before undertaking more complex tasks or attempting a task without help.

In the practice narratives, many comments (like those that follow) invoked legitimate peripheral participation, explaining how collaboration at work gave the respondent essential space and time to learn the practice.

> My qualification was that I could write and I had a good work reputation with the Tech Pubs team. I got excellent [on-the-job training] from wonderful mentors who taught me the ropes for about two years, before I felt like I knew what I was doing.
>
> I learned an enormous amount about … enterprise-level software from monitoring what developers and testers did all day, what work was assigned to them etc. I kept a close eye on the project management side of things so I always knew who was doing what and who I needed to target for information. I learned that relationship building is key to success.

One individual explained that he learned from others' mistakes:

> I … have learned the most from unpicking mistakes others have made, usually due to a lack of longer term planning and foresight.

Communities of practice are important throughout our careers. Any professional practice changes constantly because of changes in technology and tools, work processes, laws, organizational requirements, personnel, and many additional (sometimes core and sometimes peripheral) influences. Therefore, as professionals, we are all constantly learning from our colleagues.

Certification in Technical Communication

Many people choose technical communication as a second or subsequent career. If you are in this position, you probably do not have technical communication academic qualifications. Whereas some new entrants, like June in the vignette, undertake a formal academic program, others find that certification is a good alternative. Certification programs enable you to get recognition for your existing relevant skills, to develop new skills and competencies, and to improve your job opportunities. Certification also enables employers to vet for candidates who have achieved a recognized standard. When you are certified, you have a formal acknowledgement of your knowledge and skills relevant to technical communication. Your colleagues in other disciplines can understand and recognize your skills. For the profession, certification builds recognition for, and agreement about, the skill set and knowledge base. Certification sets expectations about what practitioners can do.

> Certification is also strategically important for the profession of technical communication. According to Savage, "[m]any people in the field believe we will not be able to achieve professional autonomy unless we are able to require certification for practitioners" (p. 2).[9]

The STC and tekom Europe have both developed certification programs. One respondent to the practice narratives has STC certification, and another expressed interest in this program. Although two narratives mentioned tekom Europe training, no respondent said that they had taken this certification program.

The STC Certification Model

After many attempts to develop a certification program in technical communication, the STC proposed a model in 2010. They began certifying technical communicators in May 2011 with the Certified Professional Technical Communication (CPTC) program.[32] The program was relaunched with the same name in December 2015.[33] It has three levels: Foundation, Practitioner, and Expert.[8] The Foundation level covers nine subjects:

- Project planning.
- Project analysis.
- Content development.
- Organizational design.
- Written communication.
- Review and editing.
- Visual communication.
- Content management.
- Production and delivery.

The STC assigns a textbook and a study guide. You can study independently or with support from a training body. At the end of the study period, you complete an exam.[34] Almost 300 practitioners have achieved Foundation-level CPTC certification.[35] "CPTC-certified," or variations of that term, do not appear in many job advertisements at present.

The Tekom Europe Certification Model

Tekom Europe offers a certification program at two levels: Professional and Expert. This program is standards-based, and it is supported by a competence framework that tekom Europe also developed. If you take this certification program, you begin with a one-to-one qualification consultation. This consultation is like an interview, and it helps the trainers to design your training plan. You then undertake a flexible training program: "in German or English, full-time or part-time, as academic or professional training, as e-learning or as face-to-face events."[36] Finally, to get certified, you complete examinations.

Over 1,500 candidates have completed tekom Europe's certification programs. They are based in "Germany and abroad," the exam is available in German and English, and the certification is recognized internationally in industry.[37]

Certification Limitations

> Unlike academic qualifications, which do not have an expiry date, certification by both STC and tekom Europe is time-limited.

In the tekom model, you must apply to retain certification after five years. There is no application fee for retention, but you pay an administrative fee for the certificate. The continual training and courses you need to take to retain certification are more costly. If you are STC-certified, you must gain 12 continual PD credits every two years to retain your certification.[38]

Education and Training Projects in Technical Communication

Two recent projects have developed education and training resources for this profession. These are the STC-supported Technical Communication Body of Knowledge (TCBOK) and the European Union (EU)-funded TecCOMFrame project. Outputs from both projects are publicly available. Whether you are a student, you want to move into technical communication, or you are already working, these projects will give you a better understanding of education and training in this profession.

Technical Communication Body of Knowledge

A "body of knowledge" is the core content that practitioners need to know to practice in a profession. The TCBOK project began in 2008.[39]

> The TCBOK is a publicly available resource for technical communication content.

Members of the technical communication community can contribute and share their knowledge. The TCBOK includes major sections on:

- Technical communication.
- Career management.

- Producing information.
- Researching.

A typical entry has detailed information about the topic, often with links to related content or to external sources, and references to support information and encourage further reading.

A feature of the TCBOK is its personas, over 20 imaginary characters who represent typical users of the resource. These users include students, academics, and practicing technical communicators at various levels as well as practitioners in related disciplines like translation and engineering. When the TCBOK's content is more complete, it should replace the textbook as the primary source of content for the STC's certification program.[40] The TCBOK is supported by a committee and volunteer contributors. It is constantly updatable, and potential contributors receive quite detailed guidelines. This resource has exciting potential to be a comprehensive, useful, and reliable resource for teachers, students, and practitioners. As with any such project, it needs contributions from volunteers to fulfil that potential. The home page shows an extensive list of topics that need to be developed.

> You can submit a contribution if you have expertise in any of the areas where content is currently needed.

The TecCOMFrame Project

As you saw in Table 2.2, although many technical communication programs are offered at all levels in North America, fewer programs are available in other regions of the world. In many parts of Europe, tekom Europe identified both a skills gap and an education gap that the TecCOMFrame project was developed to address. Nine partners were involved, eight from universities in Europe, and tekom Europe coordinated the project. I was one of the academic partners in this project. It ran from 2015 to 2018, and the outputs are available on the TecCOMFrame website.[41] The project built on an earlier tekom Europe project that developed a competence framework and corresponding learning outcomes for tekom's certification program.

The TecCOMFrame project team produced an academic competence framework that categorized various types of content that a higher-education program in technical communication might cover. The competence framework has six dimensions:

- Academic perspective.
- Communication and culture.
- Content.

- Management.
- Technology and media.
- Transversal competencies.

Each dimension has several subjects. For every subject, the framework includes a brief description of the background, application, scope, and a set of learning outcomes. This level of detail should help a European academic, or indeed an academic in any country, to begin to develop a course or a full program.

Figure 2.2 shows dimensions and subjects in the framework.

On the basis of this framework, the project group designed several sample curricula for programs at different levels in the European Qualifications Framework[42] (a joint framework for educational levels in EU countries). The sample programs were not designed as standards but as examples of how to use the competence framework to develop programs. Figure 2.3 shows the courses (known in Europe as modules) in a sample curriculum for a master's degree in Technical Communication. This type of program would be at the same level as the program that June described in the vignette earlier in the chapter.

I developed this curriculum with a partner. We selected learning outcomes and defined credits for each module. The different numbers of credits reflect that some modules should have more depth and content than others. Individual institutions could adopt this program in part or in its entirety and change content and credits on the basis of their program's focus.

Figure 2.2 Tekom Europe, The TecCOMFrame Academic Competence Framework, at https://www.teccom-frame.eu/competence-framework/overview/ (accessed November 13, 2020)

tekom ▆ **The Master's Curriculum**

Module name	Credits
1. Communication and language	10
2. Regulations and standards	5
3. Content	15
4. Visualisation and evaluation	5
5. Tools and delivery	10
6. Management	5
7. Training and e-learning	5
8. Research and philosophy	5
9. Internship or apprenticeship (with final presentation) or dissertation or project	30

Figure 2.3 A prototype curriculum at the master's level

> Through the academic competence framework and the sample curricula, TecCOMFrame helps educators in Europe but also worldwide to develop new higher-education programs at various levels – undergraduate and postgraduate and from a single module (or course) to a complete program – and with different specializations (e.g., linguistics, management, IT, and engineering). This project can support the development of technical communication education worldwide by enabling program development in regions where no programs are available.

The outputs from this project, like those of the TCBOK, need to be maintained to remain useful into the future. Tekom Europe has appointed an Advisory Board for Professional Development and Training; part of this group's remit is to maintain the project outputs. This Advisory Board recently launched an international network to support university teachers of technical communication.

Technical communication education, training, and certification opportunities are growing all the time. Although these opportunities are most advanced in the US, more programs are emerging in Europe and, albeit more slowly, in other regions. Many educational developments have been enabled by input and support from professional associations.

Summary of Chapter 2

This chapter has explored education and training in technical communication.

- Until recently, it was common to work as a technical communicator without having academic qualifications in the discipline. This is still true in many regions, especially in countries where academic programs are scarce or do not exist.

- Technical communication is now an established academic discipline in North America and in some countries in Europe. Numerous programs in North America, together with the STC's certification program, have increased the number of practitioners who have qualified in this profession over the past two decades. Tekom Europe also offers a certification route for practitioners.
- The vignette in this chapter describes one route among many possibilities for entry into this profession.
- Some programs are available in Asia, particularly in China. In many other parts of the world, educational development has been limited.
- Program titles, like job titles, are changing to reflect the broad range of skills now required in technical communication roles. The program spotlight outlines how technical communication programs evolved over 25 years at the University of Applied Sciences in Karlsruhe, Germany.
- In addition to undertaking academic training, practitioners need to keep their skills updated through learning on the job, self-training, and various types of PD and workplace training courses.
- Although there is no standard technical communication curriculum, projects like the STC's TCBOK and the EU-funded TecCOMFrame project have helped to identify the key competencies required in an academic program and in industry.

Discussion Questions

1. Conduct a search online for "technical communication program". Can you find a relevant program in your region? If you can, what courses are included in that program?
2. Examine one program prospectus online. How does it compare with, or differ from, the content you would expect to find on a technical communication program? How does the program content compare with the programs at Karlsruhe, described in the "Program Spotlight" section?
3. Just like job titles, some program titles are changing to reflect new roles in this profession. Search for examples of relevant programs with titles other than "technical communication." Would these titles attract your interest?
4. Table 2.2 lists some examples of PD courses and opportunities in this profession. Search for other examples of PD courses.
5. In the vignette, June described her early career and educational path, and she explained how her education mapped to her career plans. Write a short account of how you became interested in technical communication and how that interest intersected with your educational experiences.
6. June described several ways that she learned on the job. Highlight all the examples of training, PD, and on-the-job learning in the vignette.
7. The TCBOK home page lists areas where content is needed. Identify an area that you have expertise in from this list. Write a short contribution for this content area. Be sure to follow the style and format guidelines for contributors.

8. The TecCOMFrame competence framework lists dimensions and subjects that a technical communication program could include. Review the competence framework. Can you identify missing subjects or topics? What would you add or change to make this framework more comprehensive?

References

1 Davis, M. (2001). Shaping the future of our profession. *Technical Communication, 48*(2), 139–144.
2 Connors, R. J. (1982). The rise of technical writing instruction in America. *Journal of Technical Writing and Communication, 12*(4), 329–352.
3 Miller, C. R. (1979). A humanistic rationale for technical writing. *College English, 40*(6), 610–617. https://doi.org/10.2307/375964
4 Turner, R., & Rainey, K. T. (2004). Certification in technical communication. *Technical Communication Quarterly, 13*(2), 211–234. https://doi.org/10.1207/s15427625tcq1302_6
5 Kirkman, J. (1996). From chore to profession: how technical communication in the United Kingdom has changed over the past twenty-five years. *Journal of Technical Writing and Communication, 26*(2), 147–154. https://doi.org/10.2190/6P5V-TFAC-0XP7-FNPH
6 Alred, G. J. (2001). A review of technical communication programs outside the United States. *Journal of Business and Technical Communication, 15*(1), 111–115. https://doi.org/10.1177/105065190101500106
7 Thatcher B., & St.Amant, K. (2011). *Teaching intercultural rhetoric and technical communication: Theories, curriculum, pedagogies and practices.* Baywood.
8 Society for Technical Communication. (2020). *Become a certified professional technical communicator (CPTC).* https://www.stc.org/certification/
9 Savage, G. J. (2003). Introduction: towards professional status in technical communication. In T. Kynell-Hunt & G. J. Savage (Eds.), *Power and legitimacy in technical communication: The historical and contemporary struggle for professional status* (pp. 1–12). Baywood.
10 Melonçon, L., & Henschel, S. (2013). Current state of US undergraduate degree programs in technical and professional communication. *Technical Communication, 60*(1), 45–64.
11 Melonçon, L. (2009). Master's programs in technical communication: A current overview. *Technical Communication, 56*(2), 137–148.
12 Australian Society for Technical Communication. (2020). *Getting started: Becoming a technical communicator.* https://www.astc.org.au/getting-started
13 Wang, H. (2015). *Technical communication development in China* [Master's thesis, University of Minnesota]. University of Minnesota Digital Conservancy. http://hdl.handle.net/11299/172230
14 Pandit, M. (2011). Teaching technical communication in India. In B. Thatcher & K. St.Amant (Eds.), *Teaching intercultural rhetoric and technical communication: Theories, curriculum, pedagogies and practices* (pp. 177–189). Baywood.
15 TCBOK. (2019). *Program locations within colleges and universities.* https://www.tcbok.org/careers/academic-programs/program-location-within-the-university/
16 Evia, C., and Andersen, R. (2018, July/August). Preparing the next generation of leaders and innovators in technical communication. *Intercom, 65*(4), 23–24.
17 Karreman, J., Cleary, Y., Closs, S., Drazek, Z., Engberg, J., Ghenghea, V., Meex, B., Minacori, P., Müller, J., & Straub, D. (2018, July). TecCOMFrame: Developing prototype technical communication curricula. In *Proceedings of the 2018 IEEE International Professional Communication Conference (ProComm)* (pp. 69–73). IEEE. https://doi.org/10.1109/ProComm.2018.00025
18 Becker, L. C. (2018, July/August). Considering the technical communication graduate certificate. *Intercom, 65*(4), 14–16.
19 Saunders, L. K. (2018, July/August). Getting a master's degree in technical communication. *Intercom, 65*(4), 17–19.

20 Davy, D. (2018, July/August). Why get a PhD in technical communication? *Intercom, 65*(4), 20–21.
21 Yeats, D., & Thompson, I. (2010). Mapping technical and professional communication: A summary and survey of academic locations for programs. *Technical Communication Quarterly, 19*(3), 225–261. https://doi.org/10.1080/10572252.2010.481538
22 Carnegie, T. A., & Crane, K. (2019). Responsive curriculum change: Going beyond occupation demands. *Communication Design Quarterly Review, 6*(3), 25–31. https://doi.org/10.1145/3309578.3309581
23 Kienzler, D. (2001). Ethics, critical thinking, and professional communication pedagogy. *Technical Communication Quarterly, 10*(3), 319–339. https://doi.org/10.1207/s15427625tcq1003_5
24 Markel, M. (2001). *Ethics in technical communication: A critique and synthesis.* Ablex Publishing.
25 Colton, J. S., & Holmes, S. (2018). A social justice theory of active equality for technical communication. *Journal of Technical Writing and Communication, 48*(1), 4–30. https://doi.org/10.1177/0047281616647803
26 Tekom Europe. (2020). *Paths to the profession.* https://www.technical-communication.org/technical-writing/outline-of-technical-communication/paths-to-the-profession
27 Conklin, J. & Hayhoe, G. F. (2011). *Qualitative research in technical communication.* Routledge.
28 de Graaf, R. (2017). *Four walls and a roof: The complex nature of a simple profession.* Cambridge, MA: Harvard University Press.
29 Carliner, S., and Chen, Y. (2018, December). Professional development of technical communicators. *Intercom, 65*(8), 17–22.
30 Wenger, E. (1998). *Communities of practice: Learning, meaning, and identity.* Cambridge University Press.
31 Lave, J., & Wenger, E. (1991). *Situated learning: Legitimate peripheral participation.* Cambridge University Press.
32 Carliner, S. (2014). Technical communication. In V. Bhatia & S. Bremner (Eds.), *The Routledge handbook of language and professional communication* (pp. 99–111). Routledge.
33 TCBOK. (2019). *STC certification program.* https://www.tcbok.org/careers/stc-certification-program/
34 APM Group International. (2018). *Certified professional technical communicator (CPTC™).* https://apmg-international.com/product/cptc
35 Society for Technical Communication. (2020). *Certified professional technical communicator certificants.* March 18, 2020, from https://www.stc.org/certificants/
36 Tekom Europe. (2020). *Training opportunities.* Retrieved March 18, 2020, from https://www.technical-communication.org/technical-writing/education-and-training/training-opportunities
37 Tekom Europe. (2020). *International tekom certification as a technical writer.* https://www.technical-communication.org/technical-writing/tekom-certification/international-tekom-certification-as-a-technical-writer
38 TCBOK. (2019). *About the Technical Communication Body of Knowledge.* Retrieved from: https://www.tcbok.org/
39 Coppola, N. W. (2010). The technical communication body of knowledge initiative: An academic-practitioner partnership. *Technical Communication, 57*(1), 11–25.
40 Kunz, L. (2016, June 6). Certification for technical communicators: Will it succeed? *Leading Technical Communication.* https://larrykunz.wordpress.com/2016/06/06/certification-for-technical-communicators-will-it-succeed/
41 TecCOMFrame. (2018). *The joint academic competence framework.* https://www.teccom-frame.eu/competence-framework/overview/
42 European Commission. (2018). *The European qualifications framework: Supporting learning, work and cross-border mobility.* https://ec.europa.eu/social/BlobServlet?docId=19190&langId=en

3   Technical Communication Communities

Introduction

Professional communities can help us to learn, understand, undertake, and develop our practice. Professional community membership also helps us to become aware of ourselves as professionals and gives us a sense of professional identity and belonging. In communication professions, our communities are essential. Carolyn Miller believed that "to write, to engage in any communication, is to participate in a community" (p. 617).[1] You might belong to several communities. Sometimes, you might not even be aware of all the communities in which you participate, implicitly or explicitly.

Throughout this chapter, excerpts from the practice narratives and from other sources show how communities support technical communicators to interact, learn, work, and network. After reading the chapter, you will be able to identify relevant professional communities and the supports they may offer.

We begin this chapter by defining communities and focus on the features of professional communities. You then explore intersections of **professional identity** and community in a profession: how work shapes your identity, your communities shape your work, and you, in turn, shape your communities.

We explore these **types of community**:

- Professional associations.
- Online communities.
- Workplace communities.

The chapter examines the links between academic and professional communities and concludes by suggesting some **potential challenges** that professional communities pose.

In Appendix A, you can find links for all the communities, professional associations, and conferences listed in this chapter.

- If you are a **technical communication student**, you will be able to identify networking opportunities and choose communities to join that are relevant to your objectives.

- If you **work in industry as a technical communicator**, you will be able to reflect on how your career and work shape your professional identity and how you contribute to your profession. You will be able to identify existing and new communities that can support your work.

- If you are **considering a career change into technical communication**, you will be able to explore the resources and communities that could facilitate that transition.

What is a Community?

Communities are social structures that connect members who have a common purpose. We all need communities to support us in our work, education, and social lives. Communities may be loosely organized or quite formal. Although communities were traditionally based in a physical location, contemporary definitions recognize that they do not have to be bound by place. Central features include a sense of belonging, common interests, solidarity, a degree of shared identity, and shared norms.[2]

A **community of practice** (discussed in more detail in Chapter 7) is a community with an occupational or practice focus. Members are concerned with what they are implicitly and explicitly required to do in work, education, or everyday life: their practice. The community-of-practice model has four interdependent components: community, practice, identity, and meaning. This model sees practice within a framework of learning, community, and practitioner identity and demonstrates how, as practitioners, we learn our practice through our interactions with our colleagues.

Professional Identity and Communities

Regardless of whether we consider our occupation to be a profession, a field, a discipline, a job, a stopgap, or a stepping stone to something else, we come to identify with our work. Even if we think of our identity as personal and static, it is constantly changing and is shaped by our communities and our practice.[3] We develop our professional identity through our interactions with other people. At work, we are a version of ourselves, just as we might be another similar, or sometimes very different, version of ourselves at home or in a social setting. The communities-of-practice framework explains how identity and practice are linked (p. 152):[3]

> In a community of practice, we learn certain ways of engaging in action with other people. We develop certain expectations about how to interact, how people treat each other, and how to work together. We become who

we are by being able to play a part in the relations of engagement that constitute our community.

Five factors contribute to professional identity (p. 149):[3]

(1) **Negotiated experiences:** We define who we are by our interaction and participation in communities.

> Our experiences help us to develop our identity and to understand who we are. By extension, our professional experiences help us to develop our sense of our professional identity: who we are in a work setting.

(2) **Community membership:** We define who we are by the groups we join and in which we participate. Part of our identity is also formed by the groups we do **not** belong to, whether through choice or circumstance.

> Each group we join contributes to our understanding of who we are. If you are a member of a professional association, a work committee, a work team, an alumni network, and a LinkedIn group, each membership influences your professional identity.

(3) **Learning trajectory:** We define who we are by what we have learned and what we plan to study.

> Education and training experiences shape us. Your educational experiences, your educational qualifications, the training and professional development you have undertaken, and your plans all contribute to your professional identity.

(4) **Multi-membership:** We define who we are by the ways we integrate our various memberships and identities into a unique self.

We build a sense of our professional identity from all the communities (professional and personal) to which we belong, not any single one. Wenger explained how multi-membership shapes professional identity (pp. 158–159):[3]

> Claims processors do not form their identities entirely at work. They came to their jobs as adults or youths, having belonged to many communities of practice. Some have other jobs concurrently, some are students in community colleges, some are parents, some are church-goers; some are bar-goers; some have engrossing hobbies.

Wenger used the example of claims processors because his research was with members of this profession, but the concept of multi-membership is relevant in all professions, regardless of job title.

(5) **Connecting local and global:**We define who we are on the basis of both our immediate and wider communities.

We interact with local and global, small and large, virtual and face-to-face communities. Many factors affect how we interact with all these communities, and we probably behave a little differently in each one. The experience of being in each community shapes our identity.

For example, your communication patterns will probably differ in different communities:

- At your **local workplace**, you may have specific ways of communicating internally within the local community. You may use internal and local tools and terminology to communicate.
- When you communicate with **teammates in other locations**, you probably adjust your communication style to acknowledge that your colleagues will understand corporate (but perhaps not local) terms. You may still be able to use internal tools, or you may need to select others.
- When you communicate with **colleagues in the same profession** but in a different organization, you need to adjust your style in other ways. You can use professional (but not corporate) jargon, and you may need to select communication tools that all your audience can access.
- When you communicate with a **wider population**, several other factors become relevant. You need to make further adjustments depending on whether you want to communicate with experts or lay people, in related, adjacent, or peripheral disciplines, at a particular level of detail or seniority, and through an accessible medium.

Professional Associations

A professional association is a formal community where members have a common profession and undertake similar work. Communities develop around "typical organizations and institutional patterns: professional associations, professional schools, and self-administered codes of ethics" (pp. x–xi).[4] Membership in a professional association helps you to have a more solid grounding in your role and can help you to understand and expand your professional self. Joining and participating in a professional association can also help you to identify more strongly with your profession.

Features and Benefits of Professional Associations

Traditionally, the role of professional associations was to protect the profession and confine practice to a limited group that had recognized academic qualifications and paid-up membership.[5] Because most professions have opened up and diversified in the past couple of decades,[6] the role of the professional association has changed. Modern professional associations offer many benefits to individual members, the profession, and society. Benefits accrue to non-members as well as members.

- Professional associations support members, the professional work, and the status of the occupation. Professional associations protect and promote the work of the profession.[5]
- They set standards for work, recruitment, education, research, ethical behavior, and practice.
- They support individual members to collectively work for benefits that an individual could not hope to achieve alone, such as public support and increased status for the work.[5] In new fields of work, or fields where the work is not very well understood, this advocacy role is essential.
- They also enable work-related communities to develop, survive, and prosper.
- They broaden the scope of the profession and promote diversity.
- At the same time, professional associations also unify the profession. This can be an important function in disciplines characterized by diversity, like technical communication.
- Professional associations are often involved in developing, maintaining, and certifying codes of ethics and standards for the occupation. Professional codes of ethics and standards guide practitioners and support good practice. A predictable, ethical, and high-quality professional service benefits all of society.
- Professional associations usually offer resources that support practice, such as newsletters, blogs, directories, templates, and training materials. Some of these supports may be freely available to non-members, thus benefiting the whole profession.
- They promote, support, certify, and sometimes deliver professional education and training.
- They support research by funding and running research events and publishing professional or academic journals or both.
- They sometimes offer recruitment services.

> In summary, professional associations can "unify the practice and represent the profession to the public, to government agencies, and to the membership itself" (p. 366).[7]

Membership is usually voluntary. We join a professional association because we choose to, not because we have to. Members usually have to pay an annual subscription fee to the association. Your employer may pay membership dues for at least one professional association. If your employer pays for your membership, their reward is that your practice improves through that investment. Self-employed members may be able to include membership dues in tax returns.

Professional Associations in Technical Communication

Several professional associations exist to support and promote the work of technical communication practitioners and academics. You can choose which, if any, association to join, depending on your current employment status, your priorities (e.g., do you want to access training, resources, or publications?), and your career goals. Deciding which association(s) to join may depend on whether your employer will fund the subscription fee, which services you would benefit from, and which associations your peers have joined. Many of us belong to more than one professional association, and we may change our memberships over time with changes in our priorities or duties at work.

Practitioner Professional Associations

Industry practitioners benefit from membership in professional associations through education and professional development opportunities. Table 3.1 summarizes prominent professional associations that are partially or primarily focused on supporting practitioners. The table outlines the concentration and typical resources of each organization.

The **Society for Technical Communication (STC)** emerged from the merger of professional organizations that formed in the 1950s in the US. The STC is "the world's largest and oldest professional association dedicated to the advancement of the field of technical communication."[8] It has over 20 committees, local and student chapters throughout North America, and 12 special interest groups (SIGs) (reconfigured as communities of practice and communities of interest from 2021) in topics ranging from academic issues, to usability and user experience, to support for lone writers.[9] It publishes a quarterly journal, *Technical Communication*, and a trade magazine, *Intercom*, with eight issues annually. The association supports training and education through the Technical Communication Body of Knowledge (TCBOK) and a certification program (both discussed in Chapter 2) as well as running regular training events, webinars, and an annual conference. It also offers several practitioner resources for members, including a salary database, a jobs bank, and a list of ethical principles.

Tekom Deutschland was founded in 1978. In 2013, **tekom Europe**, the European Association for Technical Communication, was established as a separate entity. It encompasses several country organizations in Europe, India, and China and has 9,500 members,[10] making it likely the largest professional association for technical communication. Tekom Europe has links to educational institutions throughout Europe, runs a certification program and a jobs database, has working groups on terminology and standards, and offers resources and

Table 3.1 Selected professional associations in technical communication:
Practice-focused

Association	Concentration	Resources	Publications
Society for Technical Communication	Academic and industry practice	• Technical Communication Body of Knowledge • Certified Professional Technical Communicator certification program (both discussed in Chapter 2) • Regular training events and webinars • Chapters and special interest groups • Recruitment • Annual conference	• *Technical Communication* (journal) • *Intercom* (trade magazine) • Blogs • Salary database
Tekom Europe, the European Association for Technical Communication	Academic and industry practice	• Several annual conferences and meetings in various countries • Tekom certification program • Online resources • Recruitment • Country organizations	Specialist books and trade magazines in German and English
Institute of Scientific and Technical Communicators	Primarily industry practice	• Annual conference • Discussion forums • Local area groups	• *Communicator* (trade magazine) • Book series
Center for Information-Development Management	Primarily industry practice	• Several conferences annually • Webinars and other training events • Recruitment	• Reports, articles
Australian Society for Technical Communication	Primarily industry practice	• Annual conference • Online resources	• *Southern Communicator* (trade magazine)
TechCommNZ (New Zealand association)	Primarily industry practice	• Annual conference • Webinars • Online resources	• *TechCommWire* (newsletter)
Japan Technical Communicators Association (JTCA)	Primarily industry practice	• Regular symposium • Online resources	*Frontier* (official journal of the JTCA)

downloads to members. Its events include the TCWorld conference in German and English as well as several local conferences in member countries.

The **Center for Information–Development Management** was established in 1999 by JoAnn Hackos,[11] whose work on content management is highly influential. This organization supports information architecture, structured content, and content strategy through conferences, reports, and resources.

Professional associations for technical communication in Europe, Australia, New Zealand, and Asia operate with varying focuses and levels of engagement. They are very important in regions where limited educational programs exist (such as in India, Australia, and the UK).

- **STC India**, a chapter of the STC, was established in 1999. It publishes a newsletter, hosts annual conferences, and runs occasional events.
- **Technical Writers of India (TWIN)** is an affiliate of tekom Europe. It has several online communities, on LinkedIn, Facebook, and its own forum. It is run by a community of volunteers.[12]
- The Australian Society for Technical Communication was established in 2013.[13] It supports networking, training, and advocacy and publishes a trade magazine, *Southern Communicator*. It also hosts an annual conference.
- The UK-based **Institute of Scientific and Technical Communicators (ISTC)** was founded in 1972, following the amalgamation of three separate organizations. This organization runs an annual conference and publishes a trade magazine (*Communicator*) and a book series.

Additional professional associations operate in, among other regions:

- Switzerland (Tecom).
- Finland (Finnish Technical Communicators Association, or STVY).
- Italy (Com&Tec).
- New Zealand (TechCommNZ).
- Japan (Japan Technical Communicators Association).

Professional associations in fields related to technical communication[14] include the following:

- International Association of Business Communicators.
- Professional Association for Design (American Institute of Graphic Arts).
- National Communication Association.
- Usability Professionals Association.

Academic Professional Associations

Many professional associations focus on, support, and promote academic work. The STC has both a practitioner and an academic focus. It publishes a quarterly

Table 3.2 Selected professional associations in technical communication: Academic-focused

Association	Concentration	Resources	Publications
Association of Computer Machinery Special Interest Group on Design of Communication	Teaching and research	• Annual conference • Discussion list	• *Communication Design Quarterly* • Conference proceedings
Association of Teachers of Technical Writing (ATTW)	Teaching and research	• Annual conference • Bibliographies • Discussion list	• *Technical Communication Quarterly* (journal) • ATTW (Routledge) book series
Council for Programs in Scientific and Technical Communication	Academic program administration	• Annual conference • Discussion list	• *Programmatic Perspectives* (journal) • Conference proceedings
IEEE Professional Communication Society	Engineering communication: primarily academic	• Online communication resources • Annual conference	• *IEEE Transactions on Professional Communication* (journal) • IEEE/Wiley book series • Conference proceedings

academic journal, *Technical Communication*, and has academic and student chapters and SIGs. Although most of tekom Europe's conferences are industry-focused, TC World has an academic strand, and the European Academic Colloquium is an annual one-day academic conference. Additional organizations focus mostly or entirely on supporting academics and students in technical communication. Table 3.2 outlines some of these organizations.

These associations are all based in North America but many of them have an international membership. They provide various resources for academics and teachers of technical communication.

- The Association of Computer Machinery SIGDOC (Special Interest Group on Design of Communication) was founded in 1975.[15] SIGDOC publications include a journal (*Communication Design Quarterly*) and conference proceedings. This organization also has a student chapter.
- The Association of Teachers of Technical Writing (ATTW) was formed in 1973. ATTW publications include a book series, a journal (*Technical Communication Quarterly*), and a blog.

- The IEEE Professional Communication Society (PCS) publishes a book series, a journal (*IEEE Transactions on Professional Communication*), and proceedings from its annual conference. Several additional communication resources are available on the PCS website.
- The Council for Programs in Scientific and Technical Communication supports administration of academic programs in technical and scientific communication, also through a conference and proceedings, a journal (*Programmatic Perspectives*), and a blog.

Some of these organizations have active discussion forums and several committees with mostly academic memberships. Involvement in professional associations has many benefits for academics:

- We can participate in conferences and publish in journals that are managed by the professional associations.
- We can participate in the committees, organizational structures, and administration of professional associations.

This participation has cultural capital within academia and can lead to increased recognition, promotion, or better mobility nationally and internationally in the academic labor market.

Technical Communication Conferences

Like other professionals, technical communicators share their knowledge, skills, experiences, and research through conferences. Conferences tend to provide networking opportunities and often coincide with meetings of professional association committees, research initiatives (e.g., focus groups), and other meetings. Most conferences in technical communication are run by professional associations, and most have either an academic or a professional focus and limited overlaps. Table 3.3 lists prominent conferences in this profession and details, where relevant, of the organizers, concentration, and frequency. You can find links to the conference websites in Appendix A.

In addition to these local and international conferences, regional chapters and local groups organize smaller local conferences. For example, many STC chapters have annual or biannual conferences, and tekom Europe country organizations have regular conferences and meetings. Several academic and professional conferences in related disciplines are also relevant to technical communicators (e.g., business communication, composition and rhetoric, localization, usability and user experience, and computer science).

The Benefits of Belonging

You might question the value of professional associations for technical communicators in industry. How do practitioners use networks, communities, and

Table 3.3 Conferences in technical communication

Organizer	Conference title	Concentration	Frequency	Locations
Society for Technical Communication (STC)	STC Summit	Mainly industry-focused but includes academic presentations, covering a broad range of topics	Annual (May)	Various locations in the US
	Many STC chapters run annual conferences	Mainly industry-focused	Annually	Various US locations
Tekom Europe	TC World	Mainly industry-focused but includes an academic strand, covering a broad range of topics	Annual (November)	Stuttgart, Germany
	European Academic Colloquium	Academic	Every two years	Various locations in Europe
	Tekom Europe country associations run regular conferences	Industry-focused	Various frequencies	Various locations in Europe and Asia
Institute of Scientific and Technical Communicators	Technical Communication UK Conference	Industry-focused	Annual	The UK
Association of Computer Machinery	Special Interest Group on Design of Communication	Academic/user experience/design	Annual (Sept./Oct.)	Various locations in the US
IEEE Professional Communication Society	ProComm Conference	Academic/engineering communication	Annual (July)	The US, Canada, Europe

Organization	Conference	Focus	Frequency	Location
Council for Programs in Scientific and Technical Communication (CPTSC)	Conference of the CPTSC	Academic/program administration	Annual (Sept./Oct.)	Various locations in the US
Association of Teachers of Technical Writing (ATTW)	Conference of the ATTW	Academic/teaching focused	Annual (March)	Various locations in the US
LavaCon	LavaCon	Industry-focused: various aspects of content development	Annual (October)	The US and Europe
Adobe	Darwin Information Typing Architecture (DITA) World	Professional/corporate: DITA and structured content	Annual (Oct./Nov.)	Online
Center for Information-Development Management	DITA North America, DITA Europe, IDEAS, Best Practices, ConVEx	Professional: DITA, information architecture, content development, leadership	Several annual conferences	Various international locations/online
MadCap	MadWorld	Professional/corporate: focused on MadCap tools	Up to two conferences annually	Europe and the US
Write the Docs	Write the Docs Conference	Professional: Docs as code, writing communities, open source	Up to four conferences annually throughout the year	Various international locations/online

professional associations? Throughout the practice narratives, respondents mentioned professional associations and other communities when they were describing their work backgrounds, their education, their current work, and their future expectations. The excerpts in Box 3.1 demonstrate that members of professional associations value being involved. Benefits include **networking, sharing knowledge, keeping up to date,** and **developing skills**.

Box 3.1   Practice narrative excerpts about the purpose of membership in professional associations

I've also joined the ISTC, and have begun exploring the options for third-party courses they've approved.

I take advantage of STC webinars.

[I keep] up to date via lots of free webinars, some paid webinars from [the] Society of Technical Communication, [and a] workshop from the Center for Information-Development [Management].

To develop my skills, I've joined the ISTC, and I occasionally read their articles.

I used to go to Agile meetups and I go to devel[o]pment, Agile, or technical writing conferences when given a chance.

I keep my knowledge current [...] in tech comms in general by attending regular meetings with fellow authors, using forums, and attending conference.

I attend TCUK [Technical Communication UK, the ISTC-sponsored annual conference] every year (brilliant education and networking opportunity).

I am a Fellow of the ISTC and willingly comply with their CPD [continuing professional development] policy. ... I attended every ISTC-hosted conference (Technical Communication UK since it came into being) since I became self-employed (and a few before that, when they were held at weekends). I use the ISTC forum when I have a particular question and various vendors' websites and blogs too.

I develop my skills through academic work, volunteering for the Society for Technical Communication (STC) and I also attend STC summits and webinars.

When attending conferences, I ask a lot of questions. And I'm shortly going to be attending my first local meeting – one has just started in my area.

Diminishing Influence of Professional Associations

Despite their many advantages, some recent research suggests that membership in professional associations is falling. We already explored some factors that can dilute our sense of professional identity, like diverse academic backgrounds and job titles. Several technological transformations have also begun to threaten the authority of professions and to reduce the influence of professional associations.[6] Membership in professional associations is declining due to the following reasons:

- New labor models. Many of us work in new ways (e.g., in flexible self-employment) or in niche specializations. These changes in work patterns may lead us to identify less with any one professional association.
- "Nonprofessional career tracks into management" (p. 288).[6] Many technical communicators, for example, move into management when we are promoted. Professional associations may then become less relevant to our work.
- A divide between academic and practitioner concerns. Academic journals are becoming less relevant to day-to-day work.[16] Whereas some professional associations have an industry focus, others often use academic discourse and promote academic concerns that are not accessible or relevant to industry practitioners.

Interviewees in a UK study[17] named several reasons why membership is falling in the ISTC, the professional association for technical communication in the UK. These reasons include the following:

- An aging membership. Attrition is due to retirement rather than members choosing to leave the association mid-career.
- Diversity of emerging roles. Because many of us are no longer called "technical communicators," we are unaware of the benefits and relevance of a technical communication professional association.
- A shift toward online communities that are flexible and free to join. As the influence of professional associations fluctuates, new communities have flourished that may better match the expectations of a younger generation of workers who expect to be able to engage, particularly on social media, for free. Paying to join a professional association may seem like an unnecessary expense.

Another recent study had a similar finding about the STC's membership:[18]

Currently, the STC's 2015 Year in Review states that "membership hovered just below 6,000 members," showing a 76% overall decrease in membership between 2001 and 2015. One interviewee commented that "people don't want to pay dues" for information they can find online.

Online Communities in Technical Communication

While some professional associations are struggling to retain members, several online technical communication communities have emerged in the past decade. Some have flourished, others have floundered or disappeared. Most go through phases of popularity, some through phases of notoriety. They are likely to be free to join, they offer immediate support, and they are accessible for grassroots campaigns. In fact, some of these communities have developed organically from grassroots movements.

Communication Channels

Social media platforms are the most common communication channels for online communities. The practice narratives highlighted the importance of LinkedIn and Twitter, especially, for practitioners, as illustrated in the excerpts in Box 3.2.

Box 3.2   Practice narrative respondents describe how they use LinkedIn and Twitter

I mainly rely on LinkedIn for professional development.
I stay in touch with other professionals via LinkedIn.
I currently look on Twitter and stay connected with other technology resource teachers to hear what is up and coming.
Keeping current with lots of on the job learning, short online courses and following industry leaders on Twitter and LinkedIn.
I have a LinkedIn profile and I follow some technical writers on Twitter.

Several **Twitter** hashtags cover a broad range of technical communication-related topics from Twitter users worldwide. The most commonly used hashtags are #techcomm, #techwriting, and #techncialcommunication. Practitioners use these hashtags most to tag their tweets, although academics and service providers also use them.

- Practitioner tweets include individual queries and links to useful or funny content.
- The most common types of academic tweet include reports and photos from conferences and presentations, notifications about publications, and higher-education program promotion.
- Service providers regularly tweet advertisements for tools and services.

If you are interested in diving deeper into Twitter data, a Google Sheet has recorded all #techcomm tweets since November 2018.[19]

Blogs can also become community spaces, even though they are usually maintained by individuals or organizations. I found, in a study of blog posts about content management,[20] that blogs can be "disruptive" spaces and that diverse and thoughtful conversations play out in the comments sections. I also found in that study that bloggers commented on one another's blogs and had their own professional community. On Tom Johnson's blog, probably the best-known in technical communication, individual posts regularly attract ten or more comments, and the same individuals comment regularly.

Practitioner Communities

TechWhirl was one of the first online communities in technical communication.[21] This website hosts a discussion forum that has been operating since 1993. In its heyday, it was the main forum for online discussion in technical communication and had hundreds of new messages every month. TechWhirl continues to operate as a discussion forum, but many newer channels have emerged in the past decade. All TechWhirl archives are available online, and they offer insight into what has changed and what has remained constant in this profession in the past three decades. TechWhirl also hosts a blog, publishes magazine articles, and has a dedicated careers section.

The **Reddit "technical writing" community** is "for people who take the unbelievably complicated things that scientists and engineers devise and make it understandable for non-technical people."[22] It was established in 2012 and has over 8,000 members. Common post topics include the following:

- Education.
- Transitioning into technical writing from another field.
- Salaries.
- Developing skills.
- Developing portfolios.

This community may be especially helpful if you are a student or an early-career technical communicator since many of the posts are about joining and first working in the profession.

Write the Docs, founded in 2014, is an online community that supports technical communicators who "care about documentation." Its Slack channel has over 10,000 members (as of November 2020). This global community runs conferences and events and includes members in the US, Europe, and Australia. The community website[23] has a blog, a jobs board, multiple resources, and a link to a very active discussion forum, hosted in Slack. Slack channels cover announcements, job posts, meetups, and conferences, among other topics.

GitHub is an important community for many technical communicators, especially those who work in developer documentation/coding. GitHub enables members to collaborate, create and share repositories, learn and use code, contribute to open-source projects, and manage projects.

Several **LinkedIn** communities operate globally and at more local levels for technical communicators. They tend to be spaces for discussing job-related information, getting answers to queries, and sharing job advertisements. Although they include academic members, most posts are about aspects of industry practice. Global LinkedIn groups include the following:

- Technical Writer Forum (over 32,000 members). This group provides a "forum to discuss trends and issues affecting technical writers and the technical communication world."

- Technical writer (almost 17,000 members). This group uses the description: "principle of 'Give and Take' knowledge and ... a long lasting enthusiastic group who can provide solutions to various Pros and Cons pertaining to Technical Writing in all the Industries."
- Agile technical writers (almost 10,000 members). The About content of this group outlines who should join: "if you are an Agile technical writer or documentation manager or other Agile professional who would like to share and contribute to the collective wisdom of this group, welcome."

In addition, hundreds of local LinkedIn groups exist. For example, I am a member of two groups for technical communicators based in Ireland. It is likely that you will find a LinkedIn group for your local area. Although these groups tend to be for online discussions only, they can also lead to face-to-face meetings.

Several technical communication/content companies foster communities by sharing content, some for free and some only to subscribers. Here are just a few of many examples:

- **The Content Wrangler** publishes books, regular articles, and blog posts and runs webinars on themes of content strategy, content development, and content management. Its LinkedIn group has almost 15,000 members.
- **Cherryleaf** is a technical communication consultancy and recruitment company based in the UK and produces regular podcasts and publishes blog posts about technical communication topics.
- **Scriptorium** is a content strategy and localization consultancy that publishes blog posts and podcasts as well as various free resources relevant to technical communication.

All of these organizations promote their content and activities, share resources, and connect with followers on social media.

Volunteer Communities

Tech Writers without Borders helps charities and nonprofit organizations worldwide by connecting them with volunteer technical communicators.[24] Organizations can apply with details of a project, and volunteers can review and contribute to projects. This community is open to contributions from experienced, new, and trainee technical communicators. Volunteering for a project gives you the opportunity to develop your technical communication skills and learn from other volunteers while contributing to a good cause.

Season of Docs is a Google initiative that encourages collaboration between open-source and technical communication communities for the purpose of developing open-source documentation.[25] You apply to participate, and if your application is successful, you work with a team to write documentation for an assigned open-source project. You must be over 18, and you should have some technical communication experience. If you are eligible, you may be paid a stipend. The benefits of participating include gaining experience, collaborating with other technical communicators, expanding your network, and building your portfolio.

Academic Communities

People who teach technical communication as either their part-time or full-time occupation also participate in online communities.

- **Women in Technical Communication** (Women in TC) was founded in 2013. Its purpose is "to draw attention to the problems that women, specifically, face in the academy and in technical communication as a field."[26] Its resources include a blog and a mentoring bibliography. It has a Twitter account and uses the hashtag #WomeninTC.
- Tekom Europe launched an **International University Network** in 2020.[27] This network has virtual meetings and an online portal to support information sharing and discussion.
- Many professional associations have online community portals for academic communications. Members and non-members can access the portals:
  - The **Association of Teachers of Technical Writing** (ATTW) has a LinkedIn group and maintains an active discussion forum that members and non-members can access. The most common themes on this discussion forum are calls for papers/chapter/submissions of various types and teaching requests, mentoring advice, and teaching tips.
  - The **STC Academic SIG** maintains a listserv for members.
  - **nextGEN** is a listserv for writing, rhetoric, and digital media graduate students.

Teachers and students of technical communication also use channels like LinkedIn and Twitter extensively to share information within their own networks and to announce and discuss new publications, conferences, teaching and learning resources, trends, and movements.

Academic/Practitioner Communities

As we saw earlier in this chapter, professional associations like the STC and tekom Europe support both practitioner and academic communities, though often through separate initiatives, events, and publications. Indeed, academic and practitioner communities often seem pretty disconnected in technical communication, a divide that is regularly discussed in research.[28] Nevertheless, some projects and communities address both groups and may help to bridge the gap between academic work and practice.

Alumni Networks

Many of us stay in touch with and value our connections with university peers and professors. These networks may be loose personal connections or more formal, through LinkedIn or WhatsApp groups or an alumni mailing list. At the University of Limerick, for example, we maintain a mailing list that we use mainly to connect employers with graduates of our programs. One practice narrative respondent got their first job through a similar network:

> My first full-time job was as tech writer, I ended up there right after graduation because the company asked the university [for] some contacts for recruitment.

Another noted how they still attend events at their alma mater:

> I also attend some seminars at Université Paris 7 and stay in touch with other professionals via LinkedIn.

In the vignette in Chapter 2, you read about how June keeps in touch with her alumni network. Some practice narratives also described close connections with their university professors and peers.

Project Collaborations

Research projects often aim to connect industry and academia. Many agencies require funded research to have an industry focus or a societal impact (e.g., the European Commission's Horizon Europe programs[29]). Some projects in technical communication have been designed to draw industry input into educational development. The TCBOK and the TecCOMFrame project, both discussed in Chapter 2, resulted from academic-industry collaboration. Both projects were supported by professional associations.

- In the case of the STC's TCBOK, industry practitioners participated in a task force and have since contributed; it continues to be updated and is maintained by an STC committee.
- In the case of tekom Europe's TecCOMFrame project, industry practitioners reviewed the project outputs. Following the project, the tekom Europe Advisory Board for Professional Development and Training was established to maintain the TecCOMFrame project outputs. It comprises a mix of industry practitioners and university instructors to manage initiatives to help universities develop education and training offerings that match industry needs.

Crowdsource TPC

Chris Lam, of the University of North Texas, has developed the Crowdsource TPC platform for technical and professional communication resources.[30] The platform has three resource sections: for research and data, teaching, and practitioner resources. As the name suggests, the platform is a space where technical communication resources are crowdsourced and shared. Its purpose is to connect "the technical communication, professional communication, and business communication communities through open data, teaching materials, and workplace resources." In a spirit of open access, anyone can join and share resources they have developed. The platform also hosts a blog and a discussion forum. You can access the resources without an account, but you need to register to contribute resources or to post a message to the forum.

Workplace Communities

In addition to belonging to many formal and informal communities, we all form workplace connections that help to scaffold our work and support our learning. Interpersonal contact is essential in most types of work, not least in technical communication. Few of us work in isolation, and when we do, we often regret our lack of interaction, as these practice narratives illustrate:

> I've often been the sole technical writer, but I much prefer to work in a team.
> No options for in-house networking, and I've only encountered two other "communicators" in a corporation employing thousands of managers and developers. A local MeetUp group folded after three meetings.

Workplace communities can form in various ways, through teams that we join or are assigned to and also through sometimes deliberate and sometimes spontaneous channels.

As we discussed in Chapter 2, much of what we learn at work is from interacting with and observing colleagues. This process of learning by observing enables us to progress from new hire to skilled worker. Although this type of learning, known as legitimate peripheral participation,[3] is most pronounced early in our careers, we learn continually from observing others.

Have you been part of a workplace community? How did that community help you to learn your practice?

Our interactions within individual workplace communities are difficult to record because communities operate at a local level that is inaccessible to non-members. The practice narratives, published research, and social media provide insights into how workplace communities are configured and how they operate.

Team Configurations in Technical Communication

Contemporary work is often project-based and organized around teams. We may be individually responsible for one or several tasks, but the success of the project depends on effective teamwork. If you study technical communication, you will probably have worked in teams to complete coursework. Technical communicators are part of many types of teams, sometimes simultaneously. These teams become important communities that support us in our work.

Cross-functional Teams

Cross-functional teams are multidisciplinary, and members come from different functional areas of an organization, such as software development, marketing, and technical communication. Although team members have different types of expertise and are from different professional backgrounds, they all work toward a shared goal. Since the 1980s, many aspects of our work have changed, but cross-functional teams remain common. These excerpts from the practice narratives describe some cross-functional team configurations:

> I became part of a content team that contained two managers, an editor and four other writers: two at our company base in the north of England; one in London; one, Stateside. [...] Since then, the content team has expanded to encompass an additional writer and another editor, plus our own engineer and two localisation specialists.
>
> My role involves technical writing, a little HTML editing, working with developers, graphic designers, SMEs [subject-matter experts] and project managers on a cross-functional team.
>
> [I engage] with all disciplines involved in product design, development and delivery to ensure SMEs produce the relevant content.

Two respondents described how previous work in a cross-functional team with technical communicators influenced their decision to move into technical communication:

> I first discovered technical writing when I worked on a team with a technical writer. I worked fairly closely with her and got really interested in her work, and after a while I realized I wanted to do that job instead of the development side.
>
> I worked as a software engineer for 16 years, and towards the end of that, I got more interested in describing how the software worked for other engineers, and writing in general. There was a great technical writing team where I worked, and they were interesting to talk to.

Nowadays, many technical communicators are in Agile cross-functional teams. One practice narrative respondent described how Agile configurations potentially increase collaboration:

> I'm trying to work out new ideas about how a technical writing team should work, for example applying concepts from Agile approaches (like proactivity, focus on customer needs, specialization vs collaboration). I think that most technical writers still see their role as a solo-writer job, which is not up-to-date anymore with what big companies need.

Agile teams are discussed further in Chapter 4.

"Documentation" Teams

In large organizations, many technical communicators are part of a documentation team. It might be called the technical writing, publications, documentation, or information development team or something else. If you are part of a large team of members who have the same or similar job titles, you may be split into much smaller configurations for teamwork, as in this excerpt:

> I work for a software development company that has a large docs department (approx. 150 ppl), I am a senior technical writer and work with 3 other writers on my product.

You may be simultaneously a member of one or more cross-functional project teams and a documentation team and therefore part of both your project team and documentation team communities. In these excerpts, the narrators describe their collaborations with members of multiple work communities:

> Around 9.45, I head over to one of the engineering teams' desks to hover as a tourist at their daily standup. I do the same at 10, with another team. I then notify my own colleagues, via Slack, about the latest news, and whether it affects our own work. […] Every few days, I have an engineering sprint to attend. The team I'm (not officially, but effectively) embedded with is based up north, so I join via WebEx.
>
> [I] attend meetings with developers and with other TWs [technical writers] in the company. Assist support/services/product team with product knowledge. Work with UX [user experience] and dev teams to provide text for the product.

This quote from a survey that Tom Johnson ran and described on his blog[31] explains why it is valuable to be in both a documentation and a development team:

> Currently I'm in a hybrid model, which seems the most effective of all my experiences. We report up to a single tech comms manager in a department

that provides all types of support to engineering (in a hardware and software environment). But tech writers are assigned to a single business team that works on a single product line, and we sit part of the week with that team and part of the week with our fellow writers. Our assignments to business teams last years. This model keeps the tech comms function strong – style guides, standards, doc policies, tool support, etc. – while providing high value to the business teams we embed with.

Lone Writers

Although it is likely that you will work in a team or many teams, you may be the only technical communicator in your organization or the only one at your location. The advantages of working as a lone writer include control over your work, independence, and flexibility. Lone writers face challenges, too. The work can sometimes feel isolating. Even if you work on a team, they may not understand your work and its value if your teammates work in other disciplines. If you are the only technical communicator or the only person with your job title (whatever it is), you may also struggle to be promoted. There is no obvious career path since nobody else does this job. Lone writers also sometimes struggle to find a mentor who will advise them.

If you work as a lone writer, joining a professional association or an online group can help you to feel part of a professional community. The STC runs a Lone Writer SIG. One lone writer responding to the practice narratives explained its value:

> I have been a member of the Lone Writer SIG since I joined STC after a big layoff. I got another job … relatively easily, but it only lasted [a short time] …. Then I was out of work … and my association with the lone writer community helped me stay connected. … thus began a nine-year stint of being self-employed and heavily dependent on my lone writer colleagues wherever I went. For four years now, I have been employed full-time again, but I am the only editor. So once again, the [Lone Writer] SIG community is a life line for me.

Sociability at Work

In addition to belonging to the teams we are assigned to or choose to join, we socialize at work. These social interactions help us to learn, understand, and expand our practice. Sociability at work is important for learning, but socializing with our colleagues has many other benefits. While chatting about other topics, we plan, we solve problems, we challenge ourselves and our colleagues to try new approaches, we develop strategies, we develop confidence and commitment, and we feel increased job satisfaction.

Julian Orr[32] described his observations of service technicians at work in *Talking about Machines: An Ethnography of a Modern Job*. He explained how, as in most occupations, individuals made sense of work and developed skills and knowledge by talking and listening to and observing colleagues. Socializing was essential to the work:[33]

> [S]ociability wasn't simply a retreat from the loneliness of an isolating job. At these meetings, while eating, playing cribbage, and engaging in what might seem like idle gossip, the reps talked work, and talked it continuously. They posed questions, raised problems, offered solutions, constructed answers, and discussed changes in their work, the machines, or customer relations. In this way, both directly and indirectly, they kept one another up to date with what they knew, what they learned, and what they did.

Sociability that develops organically, as in Orr's research with service technicians, is likely more effective than social events that are imposed by management. One practice narrative was cynical about corporate communications they described as "pastoral":

> I'm less fond of 'pastoral' affairs intended, in a slightly cultish way, to quell employee dissent [...]; and 'fireside chats' from a rotating cast of execs, broadcasting from a different time zone, about which I remember almost nothing.

Even if you work remotely, your team will still be an important network. One practice narrative noted:

> I work remotely from a home office [with] three other team members who are peers, [and] a project manager, people manager, and content strategist.

Recent research suggests that working from home reduces opportunities both to learn from others through informal observation and to share our skills and knowledge with less experienced colleagues. Remote working can reduce our collective knowledge of work processes and may have an impact on our sense of professional identity.[34]

As more of us now work from home, organizations need to explore how remote workers can connect with colleagues in informal ways that foster communities of practice.

Challenges and Limitations of Communities

Communities are usually seen in a positive light. The term "community" suggests helpfulness, friendship, partnership, and general accord. Nevertheless, communities are not always positive, although we may expect them to be.

- Communities can foster, and at the same time cover up, **bad behavior** like bullying and discrimination. Like all groups, they are prone to **internal factions and cliques**. For example, professional associations are often "controlled by elite members, who can find themselves in bitter conflict with the general membership" (p. 19).[35]

- Social structures, including professional associations, are entities that enable organized responses.[36] These organized responses can lead to **groupthink**, where standing out is frowned upon. Therefore, members may be stifled when making individual contributions. Groupthink can lead to overdependence on one or a small number of charismatic leaders and to disconnection between a core group and other members. Professional associations and communities of practice, therefore, can **impose unnecessary controls** on individuals, leading to a negative impact on professional identity.

- **Excessive pride** in the community can lead to monopolization of knowledge: the community is seen as the exclusive domain of a body of knowledge. Excessive pride can also lead to arrogance. Members believe the community's current level of knowledge is the state of the art and nothing new needs to be learned.[37] Narcissism and a focus on individual needs, rather than team or business needs, can result.

- Communities are not naturally occurring, but socially constructed, and **never free of ideological influence**. They are "defined as much by **whom and what they exclude** as by what they contain; at times an act of exclusion may be essential to their continued cohesion" (p. 168).[38] Conflicts can occur around a potential member's suitability and competence to participate. Sometimes, conflict can result in positive outcomes, such as deeper discussions of purpose and inclusivity. Many communities, however, are inherently homogenous, and they exclude diverse members as a matter of course. For example, in many professions, women and minority ethnic individuals have been traditionally excluded, particularly from informal professional networks that could help to advance their careers.[39]

Communities of practice have some particular challenges. The term is loose, and some of the concepts are vague. For that reason, almost any work group with a common interest can claim to be a community of practice. Communities of practice may be enabled or diminished by the national and organizational cultures of members. Organizational hierarchies may also reduce their impact.[40]

Most of these challenges can be avoided or overcome when structures are transparent and democratic, leaders are accountable, and questioning is encouraged. Language is central to meaning making, and analyzing language use is vital to understanding the dynamics and interactions at play within a community of practice. Social media also have the potential to expose problems within communities and to hold leaders to account.

Summary of Chapter 3

This chapter examined formal and informal technical communication communities.

- In technical communication, several types of community exist, from formal professional associations, to communities of practice, to informal groupings.
- Professional associations and communities of practice offer support, resources, and networking opportunities to professionals.
- Membership in a community or professional association enables us to identify with our profession.
- Although membership in professional associations is decreasing, many online communities are thriving.
- Academic and practitioner communities tend not to intersect, but some professional associations (e.g., STC and tekom Europe) support both academics and practitioners.
- Many technical communicators are supported by communities at work, especially through team configurations and social activities.
- Although we usually perceive communities to be supportive, they have limitations and challenges, including internal factions, groupthink, and ideological influences.

Discussion Questions

1. List all the communities (you can think of) that you are a member of. Aim to think about this list as broadly as possible and include formal and informal communities. Organize the list into professional, educational, social, family, local, and other communities.
2. Describe three ways in which these communities shape or have shaped your identity.
3. Reflect on a team you were or are a member of, for either coursework or employment. What features of a community did the team exhibit? What did you learn from teammates?
4. Review the list of professional associations in Table 3.1. Which ones are you a member of or would you like to join?
5. Explore the list of online communities in technical communication and reflect on how you can participate in these communities. Search for other examples of online communities relevant to technical communication.
6. Consider the challenges of community membership. Which of these challenges have you experienced? How would you address these challenges in the future?

References

1 Miller, C. R. (1989). What's practical about technical writing? In B. E. Fearing & W. K. Sparrow (Eds.), *Technical writing: Theory and practice* (pp. 14–24). MLA.

2 Bradshaw, T. K. (2009). The post-place community: Contributions to the debate about the definition of community. *Community Development, 39*(1), 5–16. https://doi. org/10.1080/15575330809489738

3 Wenger, E. (1998). *Communities of practice: Learning, meaning, and identity.* Cambridge University Press.

4 Larson, M. S. (1977). *The rise of professionalism: A sociological analysis.* University of California Press.

5 Merton, R. K. The functions of the professional association. *The American Journal of Nursing, 58*(1), 50–54. https://doi.org/10.2307/3461366

6 Susskind, R., & Susskind, D. (2017). *The future of the professions: How technology will transform the work of human experts.* Oxford University Press.

7 Savage, G. J. (1999). The process and prospects for professionalizing technical communication. *Journal of Technical Writing and Communication, 29*(4), 355–381. https:// doi.org/10.2190/7GFX-A5PC-5P7R-9LHX

8 Society for Technical Communication. (2020). *About STC.* https://www.stc.org/ about-stc/

9 Society for Technical Communication. (2020). *Communities.* https://www.stc.org/ communities/

10 Tekom Europe. (2020). *About tekom Europe.* https://www.technical-communication. org/tekom/about-us/tekom-europe/about-tekom-europe

11 Center for Information-Development Management. (2020). *About.* https://www. infomanagementcenter.com/z-legacy-pages/about-cidm/

12 Technical Writers of India. (n.d.). *About us.* http://twin-india.org/About.html

13 Australian Society for Technical Communication. (2020). *About us.* https://www.astc. org.au/about-astc

14 TCBOK. (2019). *Professional organizations.* https://www.tcbok.org/professional-organizations/

15 SIGDOC. (2020). *History.* Association of Computer Machinery. http://sigdoc.acm. org/about/history/

16 Boettger, R. K., & Friess, E. (2020). Content and authorship patterns in technical communication journals (1996–2017): A quantitative content analysis. *Technical Communication, 67*(3), 4–24.

17 Cleary, Y., & McCullagh, M. (2020, July). Technical communication in the United Kingdom: The academic and professional contexts. In *Proceedings of the 2020 IEEE International Professional Communication Conference (ProComm).* IEEE. https://doi. org/10.1109/ProComm48883.2020.00007

18 Moore, L. E., & Earnshaw, Y. (2020). How to better prepare technical communication students in an evolving field: Perspectives from academic program directors, practitioners, and recent graduates. *Technical Communication, 67*(1), 63–82.

19 Crowdsource TPC (2019, October 24). *#Techcomm spreadsheet.* https://crowdsource-tpc.com/dataset/techcomm-corpus/

20 Cleary, Y. (2019, July). The discourse of component content management on technical communication practitioner blogs. In *Proceedings of the 2019 IEEE International Professional Communication Conference (ProComm).* IEEE. https://doi.org/10.1109/ ProComm.2019.00046

21 TechWhirl. (2020). *About TechWhirl's technical writer email discussion list.* http://www. techwr-l.com/about-technical-writing-discussion-groups.html

22 Reddit. (2020). *Technical writing.* https://www.reddit.com/r/technicalwriting/

23 Write the Docs (2019). *Welcome to our community.* https://www.writethedocs.org/
24 Tech Writers without Borders. (n.d.). *Welcome to tech writers without borders.* https://techwriterswithoutborders.org/
25 Season of docs. (n.d.). *Welcome to Season of Docs.* https://developers.google.com/season-of-docs
26 Women in Technical Communication (n.d.). *Mission and goals.* http://womenintechcomm.org/mission-and-goals/
27 Tekom Europe. (2020). *International university network in technical communication.* https://www.technical-communication.org/technical-writing/international-university-network-in-technical-communication
28 Boettger, R. K., & Friess, E. (2016). Academics are from Mars, practitioners are from Venus: Analyzing content alignment within technical communication forums. *Technical Communication, 63*(4), 314–327.
29 European Commission (n.d.). *Horizon Europe – the next research and innovation framework program.* https://ec.europa.eu/info/horizon-europe-next-research-and-innovation-framework-programme_en
30 Crowdsource TPC (n.d.). *Crowdsource technical and professional communication.* https://crowdsource-tpc.com/
31 Johnson, T. (2020, 26 July). Part VI: Results from the survey correlating org models and tech writer value. *I'd Rather be Writing.* https://idratherbewriting.com/blog/reflecting-seven-years-later-about-layoff-evaluating-survey-data/
32 Orr, J. (1996). *Talking about machines: An ethnography of a modern job.* Cornell University Press.
33 Brown, J. S., & Duguid, P. (2002). *The social life of information.* Harvard Business School Press.
34 Kniffin, K. M., Narayanan, J., Anseel, F., Antonakis, J., Ashford, S. P., Bakker, A. B., Bamberger, P., Bapuji, H., Bhave, D. P., Choi, V., Creary, S., Demerouti, E., Flynn, F., Gelfand, M., Greer, L, Johns, G., Kesebir, S., Klein, P. G., Lee, S., ...Vugt, M. v. (2020). COVID-19 and the workplace: Implications, issues, and insights for future research and action. *American Psychologist, 76*(1), 63–77. DOI: http://dx.doi.org/10.1037/amp0000716
35 MacDonald, K. M. (1995). *The sociology of the professions.* Sage.
36 Mead, G. H. (1967). *Mind, self, and society: From the standpoint of a social behaviorist.* University of Chicago Press.
37 Wenger, E., McDermott, R. A., & Snyder, W. (2002). *Cultivating communities of practice: A guide to managing knowledge.* Harvard Business Press.
38 Harris, S. R., & Shelswell, N. (2005). Moving beyond communities of practice in adult basic education. In D. Barton & K. Tusting (Eds.), *Beyond communities of practice: Language, power, and social context* (pp. 158–179). Cambridge University Press.
39 McGee, K. (2018). The influence of gender, and race/ethnicity on advancement in information technology (IT). *Information and Organization, 28*(1), 1–36. https://doi.org/10.1016/j.infoandorg.2017.12.001
40 Kerno, S. J. (2008). Limitations of communities of practice: A consideration of unresolved issues and difficulties in the approach. *Journal of Leadership and Organizational Studies. 15*(1), 69–78. https://doi.org/10.1177/1548051808317998

Part II

The Practice of Technical Communication

The second major section of the book focuses on identifying and discussing the commonalities of practice for professionals in various technical communication roles. This part of the book will deepen your understanding of the profession and its practice because technical communicators work in many types of workplace and undertake diverse activities depending on their roles.

Part II comprises three chapters:

- Chapter 4 explores common daily activities, materials, and processes.
- Chapter 5 discusses technical communication workplaces.
- Chapter 6 reflects on current trends and potential future developments in the profession.

The chapters in this part of the book acknowledge diversity, but they also enable you to identify common activities and typical workplaces, whether you are a student, a teacher, or a practitioner. This part of the book will also help you to plan for potential future developments, an important strategic activity in any profession.

4 Technical Communication Activities, Tools, Genres, and Artifacts

Introduction

The activities your do every day in your profession become your practice. Even if you work in technical communication, you probably do only some of the whole array of activities possible in this profession. In this chapter and the next one, you will explore activities (this chapter) and workplaces (Chapter 5) to develop your understanding of technical communication practice.

Practices are "patterns created through the bringing together of a set of activities, materials, understandings and skills" (p. 52).[1] In this chapter, you will investigate the **activities, materials, understandings**, and **skills** in technical communication practice. This is the essence of your practice.

In this chapter, you will explore these aspects of day-to-day practice:

- Core **competencies and skills**: the capacities that enable you to do the job.
- Core **activities**: common tasks that you are likely to do as a technical communicator. Ideally, these should map to your competencies and skills.
- **Tools** and technical skills: applications and technical expertise that enable you to produce artifacts.
- Typical **genres** and common **artifacts**: types of writing you do in industry and types of content you produce.

The content of this chapter, like that of the other chapters, includes excerpts from and analysis of practitioner blogs, the practice narratives, and published research. This chapter also includes a vignette, a composite "day in the life" of a technical communicator.

This chapter enables you to classify and analyze key competencies, activities, tools, genres, and artifacts.

- If you are a **technical communication student**, this chapter will help you to recognize the competencies and skills you need, the activities you will undertake, the tools you will use, and the artifacts you will produce. It will also help you to relate the content of your study program to your future industry roles.
- If you are a **technical communicator in industry**, you will be able to compare your activities, tools, and artifacts with those described in the chapter. The chapter will also enable you to plan professional development opportunities to match common workplace competencies.
- If you are **considering a career change into technical communication**, you will be able to identify relevant activities you already undertake, skills you need to develop, and tools you need to learn to use. You will be able to determine whether the array of skills matches your interests and aptitudes.

Competencies and Skills

In any profession, certain competencies and skills are essential. Technical communication requires a set of skills that enable us, as outlined in Chapter 1, **to communicate, explain, and help people to learn and understand, use, and interact with technical concepts or technologies.** The competencies and skills we need change over time. The terms "competencies" and "skills" are often used interchangeably, but they do have precise meanings that distinguish them.

A competence is a capacity, such as problem solving, time management, or analytic ability, that you need to be able to do a job well.

Henze explains that competence is "more than technical skill, content knowledge, or error prevention; it is also a matter of awareness, good timing, and judgement" (p. 345).[2] A competence is more general, and sometimes of a higher order, than a skill.

A skill is a more specific ability, such as the ability to use a tool, to code, or to write in a particular style.

Research and Projects about Competencies and Skills

Although there is no definitive list of competencies and skills that enable you to practice as a technical communicator, international research projects, professional association resources, and academic studies have collated many competencies and skills over the past two decades. These resources and publications do not differentiate between competencies and skills.

- The Technical Communication Body of Knowledge (TCBOK) (2010 – present) identifies skills needed for various careers in technical communication.[3]
- The German professional association, tekom, developed a cross-industry competence framework for technical communicators.[4] As the name suggests, this framework includes general and specific competencies that technical communicators need, regardless of the industry sector. Tekom uses the framework as an industry training tool.
- The European Union (EU)-funded TecCOMFrame project produced a competence framework to help technical communication instructors to determine what subjects to teach.[5] This framework covers six dimensions and 22 subjects. It is discussed in more detail in Chapter 2.
- The website of the Institute of Scientific and Technical Communicators,[6] a UK-based professional association, provides several resources that describe technical communication skills and competencies.
- Rainey, Turner, and Dayton[7] conducted an influential survey of managers to identify technical communication competencies they sought in recruits.
- Cargile-Cook, Cook, Minson, and Wilson[8] reviewed and summarized publications discussing technical communication skills and competencies.
- Brumberger and Lauer[9] analyzed job advertisements and distilled skills and competencies most often sought by employers.
- Kimball[10] surveyed managers about a range of training and education needs, including skills and competencies.
- Stanton[11] compared the skills that recruiters and hiring managers prioritize with the skills advertised in a selection of academic programs.
- Lainer[12] surveyed over 200 technical communication professionals about workplace practices. His findings include a discussion of competencies and skills.

In addition, various industry blogs, including Tom Johnson's *I'd Rather be Writing*, Larry Kunz's *Leading Technical Communication*, and blogs from technical communication service providers like Scriptorium and Cherryleaf, regularly discuss skills needed by technical communicators in different industry sectors and regions.

Frequently Cited Competencies and Skills

Table 4.1 summarizes competencies and skills that these various sources identify. They are not listed in any particular order; in keeping with the sources, I have not distinguished between competencies and skills.

Table 4.1 Competencies and skills required in technical communication

Competencies and Skills	Explanation
Communication	**The ability to question, explain, discuss, and convey information for various audiences and in multiple delivery modes.** Many types of communication skills are essential in technical communication roles, including **interpersonal** and **intercultural** communication and **collaboration.**
Information development	**The ability to produce content** that is suitable for its purpose and usable for the intended audience.
Research	**The ability to gather and analyze information to help you develop, evaluate, and manage content.**
Writing and rhetoric	**The ability to use language clearly, correctly, and precisely in a style that is appropriate** for the user, the genre, the medium, and the product. In some roles, you may need to work in more than one language.
Visual design	**The ability to select visual content to convey or support a message.**
Information design	**The ability to use design features** like color, space, and typography **and design principles** like alignment, balance, and contrast to make an effective communication.
Revision and review	**The ability to review, revise, and suggest modifications to content** to ensure it is correct and appropriate for the audience, the medium, and the context.
Technological literacy	**The ability to learn and to select, use, and evaluate technology for content development and management.**
Management	**The ability to plan, organize, and steer various elements of your work, including content, people, projects, quality, and tasks.**
Audience analysis	**The ability to determine, evaluate, and respond to the needs of content users.**
Usability and user experience	**The ability to gather, evaluate, and use information about how people interact with products, services, or texts.**
Legal and standards knowledge	**The ability to develop content that complies with laws (e.g., intellectual property laws) and standards (e.g., the Machinery Directive).**
Content strategy and information architecture	**The ability to plan, develop, structure, and manage content throughout its life cycle.**
Domain knowledge	**The ability to understand the technology/sector/ application area** for which you produce content.

This array of competencies and skills might seem intimidating, and this list is not exhaustive! If you examine the list again, however, you will probably find that you have already developed at least some of these competencies and skills through your education and life experience. It is also true that in any occupation, most of us excel in some areas but have to try harder in others.

Hayhoe distilled the most essential technical communication skills, regardless of the role (p. 151):[13]

> writing, editing, visual communication, multimedia, document design, audience and task analysis, usability testing of products and documents, and interpersonal communication.

He also considered **domain knowledge** and the **ability to use software** to be essential skills.

Transferable Skills

To succeed in technical communication, you will also need to develop some transferable skills. These are general skills needed in many roles. Brumberger and Lauer call these "more abstract than professional competencies" (p. 237).[9]

- **Critical thinking:** You need to be able to evaluate information objectively and to judge which content to include in a communication. For example, you may need to differentiate and select suitable information sources, content, media, platforms, and tools. Critical thinking featured in between 30% and 50% of job advertisements across a range of technical communication job titles in Brumberger and Lauer's study.[9]
- **Multitasking/Task orientation:** Regardless of your role, you may have to work on more than one project at any given time. In almost half of the job advertisements in Stanton's study,[11] multitasking was a required skill. It is important to be able to switch between projects and to prioritize tasks depending on their urgency and on the availability of information.
- **Organization:** You will also need strong organization skills to keep track of your time, projects, issues, updates, schedule changes, meetings, file versions, data and metadata, and content. The related skill of time management is essential in most roles.
- **Detail orientation:** Because technical content must be precise, accurate, and correct, you will need to be fastidious. In both Brumberger and Lauer's and Stanton's studies, more than one third of job advertisements sought candidates who were detail-oriented.
- **Problem solving:** Whenever we create a text, we have to solve the problem of ensuring that the content is right for the situation, the readers, and the message.[14] An aptitude for solving problems is likely to be a competence that you develop as you build experience. Problem solving was required at higher experience levels in job advertisements.[9] Johnson-Eilola and Selber (p. 3)[15] explained why they used problem solving as an organizing principle for their textbook:

> 'Problem solver' [...] acknowledges the extent to which technical communicators contribute to the development and use of technology.

- **Creativity**: Although technical communication may not seem to be a creative discipline, many studies report the value of a creative approach. Tom Johnson has written several blog posts about creativity in technical communication. As social media "stories" have begun to influence the presentation of some types of content (such as marketing communications), narrative and storytelling, and the related skill of creativity, are becoming more prominent in user documentation roles.

Important transferable skills for technical communicators include detail orientation, multitasking and organization skills, critical thinking, creativity, and problem solving.

Core Activities

In your practice, you transform competencies and skills into activities. Many of these competencies and skills map to the core tasks that technical communicators undertake. This section examines competencies and skills in terms of work activities.

Communication

Communication skills are foundational for any technical communication role. Communication is a general activity that incorporates many additional tasks, including interpersonal communication, collaboration, and intercultural communication.

Interpersonal communication is the generalized ability to interact successfully with others. Throughout your life, you have developed interpersonal communication skills. In the workplace, you need to use these skills to show empathy while being able to get your point across. Practice narratives acknowledged this requirement:

> What I underestimated previously was the importance of soft skills, communication on a personal level.
>
> I'm damn good at what I do, but I keep learning in the area of people skills... I have to be confident to gain others' respect and appreciation for my unique skills.

Collaboration is a specific type of interpersonal communication that refers to interacting with colleagues to work on tasks. Because most corporate tasks are organized around teams, collaborating is an essential task in most professions, and technical communication is no different. In addition, many teams operate "virtually" (where members are not based in the same location). This type of team configuration requires more communication for successful collaboration.

In one study, managers considered collaboration the most important competence in technical communication.[7] They identified two types of collaboration: with teammates and with subject-matter experts (SMEs).

Many practice narratives referred to collaboration with **teammates** as a central work activity. The respondents reported constant collaboration, whether through instant messages, email, a shared project management tool, or daily meetings, including scrum meetings where the team was Agile or used some Agile principles. One respondent who had previously worked as the only technical communicator in their organization appreciated being part of a writing team because of the opportunity for more collaboration:

> I've often been the sole technical writer, but I much prefer to work in a team, which I'm doing in my current job. I find the expectations are more realistic if there are other writers there. And it is also great to have peer reviews to break the isolation.

By contrast, another respondent explained their preference for working as a lone writer:

> when I had the first job [...], I realized two things: I like being a Lone Writer and I saw the key to my success in showing proficiency in many disparate fields.

Collaboration with SMEs is an essential activity in most technical communication roles. You will have to interview SMEs to learn about features, to find out about project changes, and to ensure that your content is correct. This type of collaboration may demand a more sensitive approach than collaboration with teammates. This practice narrative gave useful communication advice:

> The main thing with working with engineers is to regard their time as precious. Follow their personal style for passing on information, and get the information elsewhere if possible, so that they only need to spend time telling you something you can't get elsewhere.

Intercultural communication is an essential activity. As workplaces become increasingly globalized, it is likely you will work in international teams, made up of individuals and SMEs from different cultures, to produce products that will be translated or that must cater for multiple international audiences. Kirk St. Amant (p. 479)[16] explained the many ways in which technical communicators use intercultural communication skills:

> In some cases, [technical communicators] design web interfaces or develop online content for international audiences. In other cases, they draft documents that will be translated for release into international markets. And in

yet other instances, technical communicators work with overseas colleagues to produce informational products.

These are just some of the range of international activities in which you are likely to engage. At the most basic level of communication, you will have to become accustomed to collaborating with colleagues in different time zones, as one practitioner (based in Europe) explained:

> I collaborate with two technical authors in North America, and my development team are primarily in India. So my morning is usually busy with emails from India, then there is a lull, and then by 3–4 p.m. North America will start. The difference in time is just something that you get used to.

Intercultural communication is, of course, much more complicated than simply adjusting to time differences, and it can be fraught with misunderstandings. The iceberg model of intercultural communication[17] explains how, just as we see only a small fraction of an iceberg above the water surface, most of our cultural characteristics are below the surface and not obvious. Obvious cultural characteristics include language, holidays, clothing choices, time differences, and customs. Less obvious characteristics include different expectations about authority, time, and levels of detail in a message.[18] It is important to be sensitive to other cultures in all your communications.

Owing to a lack of nonverbal cues, **online interactions** can sometimes seem terse, even when you are communicating with individuals from your cultural background. You need to be even more careful when interacting online with colleagues and clients from other cultures.

Information / Content Development

The central aspect of your work in technical communication is developing information, whether for developers or end users. Indeed, "information developer" and "content developer" are widely used job titles in this profession. According to Kimball (p. 139),[10] the qualities that managers most value in technical communicators include "traditional content skills and their thoughtful application." Some of the most common types of content are outlined in the "Common artifacts" section of this chapter. Information development is not a single activity but a set of processes that are integrated and iterative. These processes include, among other activities,

- Research.
- Writing.
- Visual and information design.
- Revision and review.

Research

When you read about research in technical communication, the focus may be on academic research methods. Technical communicators in industry also have to conduct various types of research to produce good content. Some of the research methods you will use in your practice have parallels with academic research. They include web-based research, evaluations of competitors' products, interviews with developers and SMEs, and user research. You might need to conduct the following types of research to improve your understanding of your product and its users:

- **Library and web-based research** to learn about related products or services, including those of your competitors, and to understand processes, procedures, products, and alternates. This background information enables you to write conceptual topics, web content, and various types of background information. It also gives you a better understanding of the content that users need.
- **Tools, technology, procedure, apparatus, or policy research**. You cannot write instructions or explanations for products that you do not know how to use or procedures that you do not know how to carry out. Practice narratives mentioned the importance of having time to learn about the domain or "play with the product." You do not have to be an expert, and it is often helpful to learn as you write and develop content. You can more easily put yourself in the position of a novice or non-expert user.
- **Interviews with SMEs**. Many studies include **interviewing** among the main competencies that technical communicators require. The practice narratives indicated that interviewing SMEs is an essential, but not a straightforward, activity. It requires a great deal of communication acumen to accommodate the availability and understanding of SMEs, as these excerpts show:

  [I update] topics with SME reviews/comments, or more often than not, re-request reviews from PMs [project managers] or SMEs.
  It's imperative to get technical information from SMEs, information from product managers, branding information from marketing, etc.
  The rest of the time, I switch between one task and another as needs dictate (deadlines, the availability (or not) of test environments for screenshots or of SMEs.

- **Needs assessment**. A needs assessment involves learning about the subject matter, cataloging what content already exists, and conducting research with potential users, SMEs, or other stakeholders to identify whether a need for content exists and what type of content solution and delivery is most suitable. Needs assessment methodologies help content developers to determine content gaps and to plan and design content.
- **User observations and user experience research**. This type of research involves gathering information from users, through surveys, observation, interviews, data analysis, online user research, and other methods, and using

that information to improve our content. Increasingly, usability, user experience, and user-centered design are part of the technical communicator's job.

- **Online user research.** Thanks to the interactive features of social media tools and websites, we have access to several types of user feedback, such as user comments on social media and on the product's discussion forums, user-generated content about the product (e.g., YouTube videos that users have made to explain a feature), feedback submitted to knowledge bases, blogs, and blog comments. We can encourage this feedback and explore and use the social web to improve user assistance content.[19] We also have access to various types of usage data that enable us to understand how users interact with our content. The type of online user research you conduct depends on your role, the product, and the stage of development of the product or process.

In almost any technical communication role, you will need to undertake various research activities.

Writing

Traditional technical communication in industry involved developing instructions and procedural information, and a significant portion of the role involved writing. The most common job title reflected that skill: "technical writer." Throughout the 1980s and 1990s, guidance about writing focused on rhetorical principles, particularly principles for writing procedural texts. Some of these principles prevail in current textbooks about technical communication.

These guidelines typically advise you to use the active voice and to write clear prose and shorter sentences and to use few clauses per sentence. They can increase your text's readability and clarity.

Writing styles changed to accommodate online delivery media. A major change was the need to reduce the amount of text because of slower reading times online. Redish[20] explained how to reduce the number of words in online texts while still getting the message across. Her tips include the following:

- Writing meaningful headlines.
- Using lists rather than long sentences.
- Highlighting key words.

The **Plain English Movement** also led to the adoption of a more accessible writing style in many industry and government sectors globally.[21] In many writing situations, technical communicators have to write in restrictive ways. You may have to use templates or use a simplified version of English (or the language you are working in). For example, writers of maintenance manuals may use Simplified Technical English, a controlled version of English that is easier to read and understand, especially for non-native English speakers.[22]

Another writing development has been a shift toward **structured, topic-based** content. Structured writing evolved from Robert Horn's[23] information mapping concept. In structured writing, parts of the content (e.g., title, headings, body text, lists, and references) are identified explicitly. When you write topics (rather than long-form documents like manuals), you need to be able to chunk content and to identify topic types. If you work in a topic-based writing context, you may write using standards that constrain your writing style in various ways and that impose structure. The best-known system for implementing structured content is DITA (Darwin Information Typing Architecture), where topics are categorized into types. The three most common topic types are concept, task, and reference. DITA has many advantages, including its focus on topics and content structure and its potential to enable personalization.

> Whereas some writers find constraints frustrating, others appreciate the benefits, including structure, consistency, and clarity.

Some writing roles impose other types of restrictions that are inherent to the medium (e.g., **scriptwriting** for chatbots, multimedia or podcasts, and writing social media content).

In user assistance writing, a trend is emerging toward an informal writing style and a more casual, conversational tone. You might find this change uncomfortable if you have learned to write using a more formal tone. One practice narrative respondent discussed their reluctance to embrace this style:

> Side projects include a "modern writing" effort, where we rendered all the content into conversational English (which to many of us was counter intuitive to what technical writing is).

If your main job is writing, it is helpful to know a **second language**. This knowledge trains you to use your first language more carefully and to consider possible translation issues arising from some types of usage (e.g., if you write long sentences with multiple clauses, they will be more difficult to translate). If you write text that will be translated, you also have to be aware of your terminology choices, and you may need to use a style guide or a terminology management system to ensure that you use terms consistently.

> If you develop information for international audiences or for localization or translation, you need to reduce language complexity and references to local customs or events.

Rainey, Turner, and Dayton argued that "relevant 'tools' should include language, and especially foreign language" (p. 323).[7] In many European countries, technical communicators need to have both native and foreign language skills among their competencies.

Visual and Information Design

Many technical communication programs include a course on information or visual design or both, and many technical communicators work on aspects of information and visual design in their jobs. Kimball and Hawkins[24] differentiated between design components (like space, color, and type) and design theories (like Gestalt principles, rhetorical theory, and theories of user experience). If you studied or are studying for a technical communication degree, you may have learned about design theories and about typography, color, space, and the interactions of various design features to create usable information layouts.

Karen Schriver's studies of design in the 1990s[25] demonstrated how readers react to design decisions, even when designers are not conscious of making those decisions. Schriver's studies resonate with the concept and practice of **design thinking**.

> Design thinking is "an iterative process in which we seek to understand the user, challenge assumptions, and redefine problems in an attempt to identify alternative strategies and solutions that might not be instantly apparent with our initial level of understanding."[26]

As technical communication media are continually evolving, roles are also changing. You will likely develop types of content other than written text. Technical communicators in industry may work on visual and information design in the following ways:

- Deciding where visuals are needed in content.
- Explaining what types of visual will support text.
- Planning, capturing, editing, labelling, and updating screenshots.
- Planning multimedia content.
- Developing or commissioning video content.
- Editing visuals.
- Planning layouts.
- Designing templates.
- Writing and editing captions and alternative text for visuals.

In many contemporary technical communication roles in industry, writers write topics and structure is separated from content. If you work in this type of role, you may use a content management system (CMS) and have limited input to influence the visual design. You can, however, give feedback to designers on the basis of your knowledge and experience. The converse situation is also common: you may be more involved in design than writing. Two practice narratives reflected this scenario:

> I have little opportunity to contribute to the content writing process, but do design and create various multimedia including animations, audio, video and infographics.
>
> I develop technical manuals, design and present training with PowerPoint and shoot a lot of video and take lots of photos with a Nikon Camera and GoPros. When I arrived I was given training in Corel Draw and Adobe InDesign.

If you design content for international audiences, you will need to ensure that visuals and color schemes are neutral and inclusive. It is essential to test your designs with users from the target audience.

Revision and Review

Technical communication is a precise discipline. If you work in a team, your writing style must be correct, and it must also adhere to agreed corporate guidelines and standards. Poor text standards decrease the credibility of your organization and the authority of your content.

You may have to follow writing guidelines set out in a **style guide**. Technical communication teams tend to use style guides extensively to regulate writing. Some teams program their style guidelines into a tool (like Acrolinx) and then use the tool to check their text. Although automatic tools can help you to write correctly, they do not guarantee flawless text, and you will also need to check your text manually.

Revising your work and adhering to your company's style guide ensure clarity, correctness, and consistency. Another such mechanism is **peer review**. Content review was a highly formal part of the document development process when print manuals were the most common output medium. Review remains an important part of the technical communicator's work. When you complete a piece of writing, a member of the team or an SME may be assigned to review it to ensure that the content is clear, usable, and accurate. In the same way, you will review content that another team member has developed. These excerpts from the practice narratives explain how peer review is integrated into the writing processes:

> I write, then preview, then post content and assign a peer for review, or go to an engineer to technically review (if I can't do it myself). When my peer reviews [what I have written] and we iron out any kinks, my peer merges

the content into the main content. I then publish or wait until release to publish. I also do one or two peer reviews [for other colleagues] a day.

If I'm reviewing content, I'll add comments to the document (hosted in our CMS) before reassigning it to the author. If the author's in the same room, we'll sometimes discuss the feedback in-person.

Technological Dexterity

Many tools help you to produce effective content. The broad consensus from research is that graduates and new entrants to technical communication need to be proficient and flexible users of systems and applications. You also need to be enthusiastic about learning and using tools. Because of the variety of tools used in industry and the rapid pace of change, expertise in any individual tool is less important than willingness to learn. Cargile Cook et al. explain: "knowing specific technologies is less important than knowing how to learn technology" (p. 105).[8] One practice narrative respondent explained why understanding the possibilities of technology is important:

> Although being a technical communicator is about far, far more than being proficient in the tools, I think we must be more aware of what is technically possible – and what isn't.

The "Tools and Technical Skills" section below discusses the many types of tool- and technology-related activities that may help you in your work.

Management

In almost any modern job, a practitioner needs management skills, and many of your activities involve managing: projects, time, quality, and content. If you move into a management role, you also need to be able to manage people, budgets, and other resources.

Project management: Technical communicators "spend a good deal of their time managing projects in collaboration with other technical writers or in cross-functional teams" (p. 113).[27] Traditional project management tools track project activities in sequence over time. In the past decade, many organizations, especially in software development, have adopted some Agile project management methods. In many Agile teams, a "scrum" approach to teamwork is common. In a scrum team, content is developed incrementally in "sprints," or stages. During each sprint, the team divides "project work into small slices that can be done by one or two people in a short period of time" (p. 119).[27] Teams meet daily to update team members on their progress and to get advice on challenges. This excerpt describes one technical communicator's role in the daily meeting:

> These sprints can go on for up to three hours, but the guys are pretty good about discussing content-affecting tickets first. If I have a question, I'll pipe

up, but usually I'm silent. If the discussion gets very 'under-the-hood,' i.e., it's irrelevant to end users, then I'll shut off my mic and quietly work on something else, albeit with one ear cocked.

The language of Agile and scrum has infiltrated teams, even when they do not use this methodology, as reported in one practice narrative:

> We do not use any clear methodology (Agile, for example), although tons of Agile jargon is used in meetings.

Time management: In most corporate roles, you will work on several projects simultaneously; therefore, you will need to prioritize projects and track which projects you are working on for reporting purposes. Indeed, tracking your time can become a sizable activity, as one practice narrative explained:

> A lot of our work is micro-proceduralized, as in, for every writing task we do, it has a procedure we have to follow, and that may even contain a sub-procedure. [...] Logging completion of tasks in the schedule portal [is] quite time consuming. I sometimes spend more time doing the procedural tasks than actual writing.

Members of Agile teams report their progress on individual projects daily. Although many technical communicators have positive experiences in Agile teams, moving to Agile can be challenging from a time-management perspective, as this practice narrative explained:

> We recently adopted the Scaled Agile Framework (SAFe) methodology. Adapting the documentation to a fortnightly ...[sprint] cycle is quite challenging especially as I work on the documentation of several products simultaneously.

Quality management: The TecCOMFrame competence framework explains the purpose of quality management:

> Technical writers must understand quality management as well as quality assurance principles and strategies in order to implement appropriate quality processes and to manage projects for quality. It is strategically and professionally important for technical writers to implement quality processes and to ensure that they deliver products of high quality.

JoAnn Hackos[28] developed an influential documentation management approach in the 1990s: the information process maturity model. A key feature of this model is its emphasis on how the content development process impacts quality. Her model outlines a content development process that is sustainable and managed.

Content management: Although some technical communication content continues to be created and updated using traditional word processing and publishing tools, many organizations now use component content management (CCM) techniques. CCM is an umbrella term for the process of developing, storing, and managing pieces (or components) of content, rather than developing content in long-form documents. Effective content management reduces redundancy and increases efficiency and consistency in large content sets and across suites of content. The Center for Information-Development Management (CIDM) provides resources on content management and runs regular conferences.[29]

People management: One career path you may follow in technical communication is to move into a people management role as you progress in your career. Management activities include allocating people and resources to projects to ensure that work is completed on time, within budget, to the required standard. In this role, you will also be responsible for ensuring that work is fairly allocated and that team members are motivated and have access to training and professional development. As in many modern jobs, meetings take up a substantial portion of work time in technical communication, especially for managers. You can expect, depending on your level of seniority, to spend up to half your working week in meetings, many of them online.

Audience Analysis

Technical communication is distinctive from other types of writing because it is reader-centered. Blogger Larry Kunz[30] explained how the audience was always central to the technical communicator's activities and that did not change during his 40-year career:

> If we don't satisfy our readers, they'll go elsewhere to get information, and they might even choose our competitors' products over ours.

Technical communication must be reader-centered. This characteristic distinguishes it from other forms of communication. For example, creative writing does not have to be understandable on first reading, and marketing writing does not have to be usable. Readers of technical communication content, however, need to understand complex concepts and perform procedures accurately and efficiently.

As a technical communicator, you will need to conduct research with audiences to understand their needs. Schriver[25] described three ways to approach audience analysis:

- Classification-driven: This approach involves describing probable characteristics of the audience, like their age ranges, locations, interests, and professions.

- Intuition-driven: This approach involves imagining an audience member or members and designing content based on their expectations, needs, and interests.
- Feedback-driven: Schriver's preferred approach, feedback-driven audience analysis involves gathering direct feedback from individual audience members and using that feedback to improve content.

Although feedback-driven analysis was an expensive option (because of the time-consuming nature of interviewing or observing users), social media offers a more cost-effective platform for user feedback. In her book on audience analysis, Breuch[31] described how social media feedback can influence website development.

Content Strategy and Information Architecture

As we saw in Chapter 1, many new "content" roles are emerging within technical communication. Of these, the role that has attracted perhaps the most attention is content strategy. Although content strategy is a job title in itself, it is also a set of activities that might be your responsibility in any technical communication role. It involves tracking and planning for the development, deployment, and management of content throughout its life cycle. Scriptorium,[32] a company that specializes in content strategy consultancy, identifies several activities that contribute to content strategy, including analyzing existing content and content development approaches, identifying gaps and needs, and planning content sets based on needs assessment.

> Companies that produce large amounts of content need content strategists to enable them to manage content efficiently and to ensure that the content is efficient and consistent across platforms and delivery media.

One practice narrative respondent described their content strategy role in this way:

> [I am a] content strategist managing the end to end customer content journey, ensuring the client has the content they need no matter where they are on their journey, including Try, Buy and Use phases.

The excerpt shows how content strategy permeates all phases of content development and interacts with many other roles.

Information architecture is concerned with structuring large amounts of content to ensure that an information system is usable. An information architect may work in user experience, web design, or technical communication. In technical communication, information architects classify, label, and structure content. They use **taxonomies** to group similar types of content, they use **metadata** to label the content, and they use **rules** to ensure that content can be accessed within a system.

Domain Knowledge

Technical communicators work in many sectors, including manufacturing, information technology (IT), health care, government, and financial services. Although technical communicators do not have to be experts in the domain, some subject-matter expertise is helpful. For example, if your job involves writing application programming interface (API) or other developer documentation, you will need to be able to read code. According to Stanton's study, 38% of job advertisements sought subject-matter expertise. Larry Kunz,[30] reflecting on activities that had changed and prevailed in the profession in his 40-year career, explained that nowadays **writers choose specializations** but that, early in his career, companies expected to have to train them.

The background section of the **domain knowledge** subject in the TecCOMFrame[5] competence framework explains why domain knowledge is important:

> Technical writers mainly work for economic sectors, like machinery or software development where basic knowledge about engineering and computer science is needed to develop information products. Often, they design information about technology for professional target groups, such as engineers. They must also communicate with professionals from engineering disciplines or computer science. For this, technical writers should have a general understanding, should know about the technology, and understand its functioning.

If you are new to technical communication, you may not have identified your domain yet. It is possible to move from one domain to another (e.g., from enterprise software to financial services). You should choose your first job carefully, however, because when you develop experience, you will find it easier to progress to a new job in the same domain.

Of course, if you do not have domain knowledge, you can still learn on the job, as this excerpt from the practice narratives explains:

> I am the lone editor for Pre-K–12 curriculum in the STEM [science, technology, engineering, and mathematics] subjects of computer science, biomedical science, and engineering. I have learned A LOT about these fields during this job. I had no background in education nor these subjects.

In some domains, you will need to understand the legal and regulatory framework and apply laws and standards to the content you develop. For example, products and their documentation produced in the EU must conform to the Machinery Directive.[33] Data security and protection laws also differ from region to region. The TecCOMFrame competence framework includes laws and standards in its list of subjects. If knowledge of laws or standards is essential for your work, you may need training or other supports.

Tools and Technical Skills

Technical communicators use a vast array of tools and technical skills in their work. This section examines some of the many tools and technical skills that are referenced in the practice narratives and in blogs, online resources, and trade articles. The term "tools" encompasses the array of software and other applications that you may use in your career, depending on several factors:

- The organization you work for.
- The types of content you produce.
- The projects you work on.
- The common content development methods in your industry.

Range and Variety of Tools

As in other industries, you will use tools for many purposes: writing, content development, design, collaboration, coding, and project management. You may use open-source, in-house, and proprietary tools, often all of these in a single role.

- Open-source: The source code for the tool is available and can be modified. Open-source tools are usually free to use.
- In-house: The tool has been developed by your organization to respond to a particular need.
- Proprietary: The tool is widely used in industry and available for purchase.

These excerpts from the practice narratives show how flexible practitioners are in using tools for different tasks:

> To write XML documentation (DITA), I use Oxygen. I also use Slack, email, Zoom and face-to-face for everyday collaboration. Projects and tasks are managed in Jira.
>
> XML Mind essential for daily DITA writing; unstructured Frame-Maker for maintaining older documents; full Acrobat + Adobe CC for graphics.

This excerpt demonstrates how part of your job may involve **migrating content** when the organization adopts a new authoring tool:

> We used to write content for white papers in Word (+PDF) and we write functional help in a function embedded in the software; but we have been

moving to Madcap Flare recently (we are currently migrating all the white papers into Madcap Flare; we'll migrate functional help in the future).

This excerpt indicates the **range of tools** that one respondent uses:

Essential software includes Microsoft 365 (Teams, Office, Cloud, etc.), Filmora, Camtasia, Photoshop, Snagit, FrameMaker (becoming less essential), Moodle, Confluence, JIRA, Audacity, etc.

By contrast, other practice narratives reflect much less demand for technical dexterity. One respondent noted:

98% of my interaction is over email, 90% of what I do is in Word.

In the vignette later in this chapter, you will see the range of tools that one technical communicator uses in a single day. This range is reflective of the variety recorded in the practice narratives. These narratives suggest some additional patterns:

- XML/DITA authoring tools are widely used. At present, there is no industry standard, but Arbortext and Oxygen XML are popular. Adobe FrameMaker and MadCap Flare also support XML authoring.
- Microsoft Word is a secondary tool for many technical communicators, who use it for workplace communications and reports and to a lesser extent for product information development. Nevertheless, a CIDM 2019 trends report found that over a quarter of respondents used it as their primary tool.[29] MS Word is also the primary, or only, tool named in many job advertisements.[9]
- Because most professionals sometimes have to present their work and Microsoft PowerPoint is the industry standard, most professionals use this tool, but it is rarely a primary content development tool.
- In some sectors (particularly for developer documentation), technical communicators use editors to write documents in lightweight markup languages.

Despite these emergent patterns, a key message is that there is **no industry-standard content development tool** in technical communication. In a 2017 blog post, Tom Johnson[34] responded to a reader asking for advice about which tools to learn for an API documentation role:

The short answer is that there are many different tools for creating documentation, and there's no clear industry standard in this space.

A manager of a global technical communication team told me that members of his team, based in different locations around the globe, used 14 different authoring tools when he began managing the team. He worked in cooperation with the writers to reduce the number of authoring tools and streamline writing processes. Moore and Earnshaw[35] explored tools in interviews with practitioners. They also reported great variety, even among their small sample of interviewees (six practitioners).

Common Tools and their Functions

Although FrameMaker and RoboHelp were industry standards for producing manuals and online help content for a couple of decades, tools have become much more diverse in the past 15 years. These are some of the most common tasks for which we use tools in technical communication:

- Word processing (e.g., Microsoft Word).
- Presentation design (e.g., Microsoft PowerPoint).
- Text/code editing (e.g., Atom).
- Topic-based writing (e.g., MadCap Flare).
- Web development (e.g., Dreamweaver).
- Graphic design (e.g., Photoshop).
- Content management (e.g., Sharepoint).
- Project and time management (e.g., Slack).
- Collaboration (e.g., Microsoft Teams).
- Meetings (e.g., Zoom).
- Social media (e.g., Twitter).

Table 4.2 shows a more comprehensive list of tools and their applications, derived from the practice narratives, and these additional sources:

- Current international job advertisements (October 2020) on indeed.com and monster.com.
- A list of tools used by technical communicators identified in a 2018 census of technical communicators.[36]
- Recent academic articles that discuss tools used in technical communication.
- Blog posts, podcasts, newsletters, Twitter feeds, and LinkedIn posts discussing tools in technical communication.

Some tools combine more than one function. For example, many project management tools also have collaboration functions.

Table 4.2 represents my interpretation of how the various sources categorize and explain tools. I categorized tools according to their primary functions, but they are not listed in any particular order.

Table 4.2 Tools commonly used in technical communication

Category	Purpose	Examples
Content authoring	Multi-format publishing	MadCap Flare and Adobe FrameMaker. FrameMaker was the early industry-standard authoring tool for technical communicators. Some technical communicators use Microsoft Word as their primary authoring tool. Google Docs and LibreOffice Writer are free alternatives. Tools like Acrolinx help teams to adhere to pre-programmed style guidelines.
	XML authoring	Arbortext, XMLmind, and Oxygen XML Editor. These tools are used for editing and publishing XML documents. They support several output formats. Both Flare and FrameMaker support various publishing outputs and XML and DITA authoring.
	Help authoring	RoboHelp and MadCap Flare. As embedded help has become more common and tools like MadCap Flare offer multiple publishing outputs, including Help, the popularity of help authoring tools (HATs) has waned.
	PDF editing	Adobe Acrobat Pro.
	Presentation	Microsoft PowerPoint, Prezi, LibreOffice Impress, Google Slides, and Keynote. Presentation tools are also used for storyboarding and designing posters.
	Text editing/ coding	TextMate, Notepad ++, and Atom. Some writers use editors to write documents in AsciiDoc, Git, and Markdown.
	Wiki development	Atlassian Confluence. Alternatives include Helpie (a WordPress plugin) and Mediawiki.
	Blogging	WordPress, Blogger, Wix, and Weebly.
Design and multimedia development	Web design, web authoring, and blogging	Adobe DreamWeaver for web design and development. Several web-based alternatives are available. Among these, Weebly, WordPress, Google Sites, and Wix are popular. Jekyll enables development of static websites.
	Graphics	Adobe Photoshop, Fireworks, and Illustrator, Gimp, Corel Draw, and, for much less complex projects, Microsoft Paint. Several screenshot editors are available. Of these, Snagit is the best known. Microsoft Visio is used for creating flowcharts and other diagrams.
	Computer-aided design	Autocad and Creo. These tools are used predominantly in heavy manufacturing documentation.

(*Continued*)

Table 4.2 (Continued)

Category	Purpose	Examples
	Page design	Adobe InDesign for page design, layout, and typesetting.
	Animation	Vyond and Adobe After Effects.
	Video editing	Adobe Premier Pro, Filmora, and iMovie.
	Screencasting	Adobe Captivate, Panopto, Techsmith Camtasia, and Jing.
	Audio recording	Audacity, Wavepad, Adobe Audition, and SoundForge.
E-Learning development	Learning management systems	Moodle, Blackboard, Sakai, and Canvas.
	Course development	Articulate Storyline, Articulate Rise, Camtasia Studio, and Adobe Captivate.
Project management	Team collaboration	Microsoft Teams, Trello, Google Workplace, and Slack.
	Project tracking	Jira, Slack, Teams, and Microsoft Project.
	Videoconferencing and meetings	GoToMeeting, Skype, Zoom, WebEx, Google Meet, Slack, and Teams.
	Instant messaging	WhatsApp, Skype, and Slack.
	Content management	Ixiasoft (a DITA component content management system), SchemaST4, Kentico Cloud, GitHub, and Sharepoint.
Translation	Translation management	SDL Trados and Localize.
Social media	Marketing, user support, peer support, networking	Twitter, Instagram, Facebook, and LinkedIn.
General administration	Word processing	Microsoft Word is widely used in most corporations and in most roles. Alternatives include LibreOffice Writer and Google Docs.
	Email	Gmail, Microsoft Outlook, and Proton.
	Spreadsheets and databases	Microsoft Excel and Access, LibreOffice Calc, and Google Sheets.
	Calendar	Microsoft Outlook, Google Calendar, and iCal.

Other Technical Competencies

In addition to knowing tools, you will need to be able to read and understand, and possibly write, code. The Write the Docs[37] and Docs as Code[38] movements encourage writers to treat documentation like code. Several practice narratives reflected on a trend toward writing in **lightweight languages** like AsciiDoc and Markdown:

> Using AsciiDoc, I document features of an enterprise-level software project. I use Git to manage source control. I prioritise incorrect things in live docs while writing future features for releases.

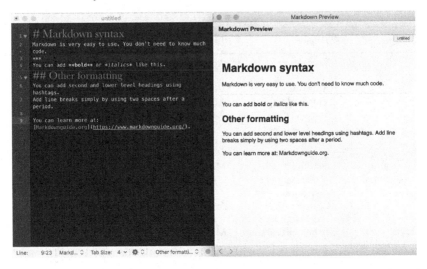

Figure 4.1 Markdown editor and preview example.

Currently I'm using Madcap Flare and AsciiDoc. Engineers prefer to use AsciiDoc, it can be stored near the code. There can be either local stylesheets to generated PDF/HTML or it can be imported as HTML into Madcap Flare which is a bit more work in both fixing the input and maintaining updates from engineers. I love both tools.

Primarily work with DITA using Oxygen XML Author. Also use Markdown and GitHub. Occasionally produce videos using Camtasia. I use some internal tools for building the content from DITA to HTML5.

These lightweight languages are easy to use and read. Figure 4.1 shows a basic example of Markdown text. The left-hand pane shows the syntax and content in an editor; the right-hand pane shows the output.

Moore and Earnshaw (p. 68)[35] cited one practitioner who included an understanding of code in their recruitment priorities:

> I am looking for someone who is not afraid to dig down into C++ or Python or Java ... if they don't understand the technology, all they're doing is copyediting [and] that's not effective writing.

The level of coding knowledge you will need depends greatly on the role. Some writers are more enthusiastic about coding than others. Only a few academic studies suggest the need for coding skills, but for some types of jobs like writing developer documentation, being able to at least read code is essential.[39]

Tools are Just Tools!

Tools and applications have been a tricky and sometimes controversial subject in this profession, for teachers and students, and in industry. Students want to learn the most current versions of the tools that are most widely adopted since you believe that knowing those will help you to get your first job. This is a reasonable expectation given that many technical communication job advertisements mention authoring/development tools.

Teachers, on the other hand, want to ensure that we are helping students to develop skills for a long career, not just for a first job. Although tools enable you to produce content, they cannot substitute for the rhetorical, design, management, and other complex skills and competencies you will need when you work as a technical communicator.

Our preoccupation with tools may be a barrier to higher status and a stronger profession. Carliner[40] argued that students need to learn longer-term "durable" skills that they will use throughout their careers, like strong writing and rhetorical skills and an understanding of design.

> When we think about the characteristics we admire in other professionals, the software they use is usually not on the list! We are more likely to identify their higher-order competencies and skills, honed over time, that enable them to make professional creations, decisions, and judgements.

Because there is no industry standard and because tools are just that (tools that get a job done, like a spade or a pen), the main aptitudes you need to demonstrate regarding tools are flexibility and adaptability. Employers want to know that you have used many different types of tool, that you know the affordances of different tools, and that you can learn new tools quickly. Being able to learn many tools is usually more important than knowing a single tool.

Vignette: A Day in the Life of a Technical Communicator

How do these skills, activities, and tools come together in practice? In the practice narratives, respondents wrote narrative accounts of a typical day. I also asked them to consider:

- Features of their workplace.
- Their role.
- Essential software and/or hardware tools and/or applications.
- Professional development activities (e.g., essential courses or recent training).
- Collaboration and involvement in networks.

In the vignette that follows, you will read about one day in a software writing role and the many activities that Oscar, a composite character, undertakes to do his job. A clear consensus from the 59 responses to this narrative question was that there is no typical day, but this vignette describes activities, tools, and skills that many respondents mentioned. The vignettes in Chapter 5 zoom in on different types of workplace.

No day is typical. I'm sure everyone says that! But I will try to describe regular activities and projects. I leave home at about 7.30 a.m. and drive to the office. My company currently does not support working from home in large numbers, though exceptions are made in some situations. I arrive early, at about 8 a.m. I like to have some time to plan my day and to use the peace and quiet to concentrate and to catch up before the office hubbub starts up. I work in an open-plan office, so it can get busy as the day progresses.

I work in a software company, on an Agile development team. We work in three-week sprints. This is a short phase where we work to complete a small portion of work. For any typical sprint, I spend some time planning with managers to ensure our team has enough resources to do the work, some time planning my content and workload alone, some time interviewing subject-matter experts, some time using the software, some time writing, some time taking screenshots or adapting graphics to get the visuals I need, some time reviewing work, and some time making changes based on review comments.

The developers use a Jira ticketing system to track and communicate about our projects, and our team uses Slack to collaborate and manage our projects. I work in a multi-site, multidisciplinary team, with members in three countries, so there will always be updates from other time zones, where work finished later or began earlier than for me. I check our Slack workspace and also my email for updates. I'll also check the updates in my Jira 'queue.' Often these updates need some kind of action, so I may have to write responses or forward the queries if I cannot deal with them myself.

I work on several projects simultaneously, so I prioritize Jira issues depending on their urgency and deadlines, but to an extent I'm able to decide what I want to work on and to control my own work day. Although my job title is Technical Writer, I spend less than half of any day writing and quite a lot of time on research, communication, various meetings, some training, some administration. There are a lot of corporate reporting activities that I've gotten used to scheduling into my day.

I do most of my writing in the morning. If I'm writing content from scratch about a new product feature or an enhancement, I'll view the Jira issue for this feature/enhancement. This is a note in Jira that tells me what the feature is or does. I'll then use the acceptance criteria (the requirements set out by the developer) as a basis for drafting my content. If I can access the feature/enhancement to test it on a development environment, I'll see how it works and, if possible, I'll take a screenshot that shows how it's used.

We use Oxygen XML Author to write DITA topics. We follow a published style guide to ensure we are consistent in our approaches to writing, spelling, grammar conventions, terms, and so on. I use Microsoft Word for any internal documents I have to write (reports, newsletters, and so on). I occasionally have to give presentations, and I use PowerPoint to create these. For one (legacy) product, we use FrameMaker to update a user guide. Updates are needed if we have to document new features, change screenshots, or fix bugs. I use Snagit to edit screenshots.

We have a peer-review process where we're assigned work from other team members to review. The style guide also helps when reviewing peers' writing. If I'm reviewing content, I'll add comments to the document (hosted on our content management system) before reassigning it to the writer. If the writer is based on my site, we'll sometimes arrange to meet to discuss the feedback together. I also have to revise my content after peer review. I don't take the reviews personally. The revisions are usually about lack of clarity or about style guide conformance, and I almost always understand and agree with the reviewer.

I take coffee and a snack to work with me, so I don't need to take a morning break. I do take a lunch break though, for about 45 minutes. Most days I go to the canteen and eat lunch with my colleagues. The canteen is always a sociable place, but we mostly talk about work because it's the main thing we all have in common. These conversations can be very helpful. I work out ways to deal with challenges, or discover that other people have the same challenges, and that is consoling. That would be typical, but at least once a week, I go outside for a walk during my lunch break.

I spend a sizable proportion of each day either in formal or informal meetings. Because I work on a team with members in other countries (and time zones), we don't have a daily standup meeting, but other teams I've been on have these every morning. Our team uses Slack to share files, chat and for a weekly synchronous team meeting. On any given day, I also have a few meetings with my manager, SMEs, or other writers on site, for other information gathering or for other committees.

We sometimes have on-site mandatory training (e.g., on data protection, security, or new corporate policies). We occasionally have editorial workshops; these are really helpful. About once a month, I take a webinar on something related to work that I want to know more about. I occasionally write for our site's newsletter. Our site has partnered with a local homeless charity, so every few months we have a half-day volunteering activity. Obviously, that's not something that happens on a typical day, but it's nice to be involved in volunteering.

I leave work most days at about 5.30 p.m., and it's a short drive home if the traffic is light. I rarely have to work late, and I never take work home with me. Once I get home, I forget about work until tomorrow.

This vignette represents only one of so many possible days in a technical communication role. The vignettes in Chapter 5 offer different perspectives, depending on the sector, the organization, and the workplace.

Technical Communication Genres

You will create content in different styles depending on its purpose and context. These different styles, or categories, of text are known as genres. Henze defines a genre as "shared conventions of text and situation" (p. 337).[2] The situation is important in technical communication because you need to consider your audience and the settings in which they will use your text.

In the vignette, Oscar works in a software environment and the main type of writing he does is **procedural**, but he also creates some **reports** and **presentations**.

One of the best-known genre theorists, John Swales,[41] identified "moves" or characteristics of academic genres. You might be familiar with his work if you have studied composition or academic writing. In your academic work, you use typical academic genres, including essays, proposals, reports, and presentations. Each of these genres has its own conventions, or moves, and in order to complete an assignment successfully, you must adhere to those conventions. For example, a formal report will include preliminary pages such as a title page, an abstract, and a table of contents. Likewise, in technical communication, several genres are common, and each of the common genres has typical features. For example, procedural writing is a genre with a set of characteristics. You write instructions in steps and as commands (known as the imperative mood). Instructions include typical features of the genre, like images to show a device or to illustrate steps, troubleshooting tables, and safety notices.

Knowing the purpose, context, and components of a typical genre enables you to understand why the text includes the information that it does and not other information that would be equally possible to include.

"Knowing a genre means not only being aware of its form and textual features but also being aware of when the genre can be used, in which situation, for which purpose, and which content is appropriate" (p. 286).[42]

Of course, adhering to conventions is only the starting point; you must also produce clear, accurate, and precise content to ensure that your work is accessible. You might also need to deviate from or adapt typical or common genres if the conventions do not match the rhetorical situation or your readers' needs.

When groups do not share an understanding of genre, sometimes very serious and even fatal problems can occur. Edward Tufte[43] examined a presentation that engineers gave to managers in the lead-up to the Space Shuttle Columbia disaster, a fatal 2003 US space mission. He selected one slide for analysis; the slide explained the potential damage that foam could cause if it fell from part of the shuttle during launch, a fairly common occurrence. Tufte concluded that a lack of shared understanding (between the engineers and the managers) of the way the information was presented on this slide was partly responsible for the disintegration of the space shuttle on its 28th mission.

Purposes of Genre

In technical communication, some common purposes of our content include the following:

- To instruct.
- To explain.
- To provide background information.
- To assist.
- To enable.
- To encourage.
- To solve a problem.
- To provide support for a claim.
- To reassure.
- To influence.
- To prevent.
- To engage.

You can probably think of other purposes for the content you produce. Sometimes we have multiple purposes. Deciding the text's primary purpose, and other secondary purposes, enables you to choose an appropriate genre.

User Contexts and Genre

Carolyn Miller[44] believed that instead of thinking about documents as ends in themselves, we should think about them as a means to achieve social action.

We should consider this question first: What will the user want to do with this content?

The setting in which the users of a document find themselves influences genre selection. For example, some users will use your content while performing tasks, perhaps outdoors, perhaps in dangerous situations. They may be working in a loud environment where they cannot access sound. They may have limited internet access, or they may access the content on a small screen. All of these settings influence the genre that you select for your communication and ultimately the content you develop. The setting may also dictate or at least influence the artifact that you choose to produce.

Genre Conventions

We expect certain modes of expression in different genres (whether a manual, a website, a help file, an instructional video, or a training course). Some genres are more formal than others. For example, the content of a knowledge base article is likely to be presented in a serious and formal style. By contrast, social media posts are typically less formal and have more scope for humor and fewer restrictions on form and style. The style may also be personal rather than corporate. Nevertheless, social media posts have other genre restrictions such as limits on characters and typographical formatting.

A **template** is a basic sample text that often includes conventional features of a genre. For example, the templates in Microsoft Word give you some insight into the features of genres such as newsletters, brochures, cover letters, journals, resumés, and academic papers (see Figure 4.2).

Henze[2] uses the examples of a business letter and a quick start guide to demonstrate genre conventions and our expectations as text users. If you compare examples of any two genres, you can identify how words interact with other text features, like visuals and information design. Analyzing the genre can involve analyzing the features of the text or analyzing the features of the social context (the discourse community).

Figure 4.2 Microsoft Word™ sample templates. This way to microsoft.com. Used with permission from Microsoft.

Typical Genres in Technical Communication

Knowing the genres that are typical for your workplace, industry sector, and audiences enables you to select and use them appropriately. In industry, several genres are typical, and you will need to recognize the conventions of each of these genres to help you evaluate which genre to select for a specific type of communication. A genre constrains us as writers because we have to conform to its conventions. These constraints help our readers. Instead of having to learn how to use a text or what to expect from it, readers can use their knowledge of genre conventions to enable them to find and use information quickly.

> Genre constraints also help us as writers because we begin with an outline of a typical example that we can adapt.

Killingsworth and Gilbertson[45] identified three main genres in technical communication: the report, the manual, and the proposal. Their work predated the internet and the evolution of topic-based writing, which have resulted in new genres. In the DITA[46] standard, there are three information types that reflect primary instructional genres in industry:

- Task information: gives instructions on how to do something, usually in steps.
- Conceptual information: explains an idea or provides background information. This type of information should be short and simple.
- Reference information: provides source information, often in tables in a technical document.

An article about technical communication on the **Genres Across Borders** project website[47] analyzed research to identify technical communication genres:

- Technical and scientific reports.
- Proposals.
- Academic articles.
- Computer documentation.
- Email.
- Various types of medical documentation.
- Web resumes.
- Design critiques.
- Corporate promotional videos.
- Patent drawings.
- Call center communication.

In the workplace, you will use genres as a guide to help you to produce different artifacts or outputs. The next section explores common artifacts.

Common Technical Communication Artifacts

Although the terms "genre" and "artifact" are sometimes used synonymously, I have distinguished them to show that **a genre is a mode of expression of content** and **an artifact is a product**. The purpose of the distinction is to help you to consider the process of content development separately from its product. This distinction is not always necessary, however. For example, a website can be both a genre and an artifact.

Publishing Formats

Artifacts are influenced by publishing trends. For example, although manuals persist as a format, they are less likely to be published in print today than they were 20 years ago. Much of the content that you produce will be published in several formats, like HTML, PDF, and print.

The CIDM trends report[29] distinguishes several **publishing formats**:

- HTML (web content). Much of your content is published and available online, either publicly through the internet or in closed intranets that users access through work or subscriptions (also known as extranets).
- Electronic PDF. With HTML, electronic PDFs are the most common publishing format, according to this survey.
- Help and embedded user assistance.
- Print. Although print is less common, it is still a prevalent publishing format.
- Dynamic delivery (also known as "intelligent content"). Content is published online in response to a specific, contextualized user query.
- Video.
- Mobile applications.
- Wiki content.
- Learning management system content.

In the workplace, you use your knowledge of genre to select publishing formats to produce artifacts that are suitable for the audience and purpose.

> In the vignette, Oscar produces different artifacts (e.g., user guides and presentations). He also writes DITA topics that might be published in multiple artifacts, including websites.

Typical artifacts change over time as technologies and user devices change. Nevertheless, research indicates that many artifacts, including user manuals and presentations, have persisted.

- Rainey, Turner, and Dayton[7] ranked information products on the basis of how often managers reported them. The most common products at the time of their survey (2005) were PDF and hard-copy documentation, online help, style guides, online reference materials, web pages, and training materials.

- Selfe and Selfe[48] used text clouds to show common technical communication artifacts, based on ten years (1996–2006) of research articles from two journals. The artifacts they identified included documents, manuals, reports, help, presentations, and research studies.

- The most common artifacts listed in job advertisements for "technical writers," according to a 2015 study,[9] were user guides and technical documents, reports, presentations, video and multimodal content, website content, and database content. For other job categories, other types of content were prominent. For example, for content developers and marketing writers, social media content and marketing materials were the most common artifacts mentioned in job advertisements.

Manuals/User Guides

Printed user manuals were the most common artifact in technical communication in the twentieth century. Indeed, printed manuals for machinery predate the profession of technical communication. Many technical communication researchers have analyzed early printed manuals for tools and machinery. For example, Brockman examined various technical communication formats from the late eighteenth century to the 1990s.[49] When lay users with limited or no computing knowledge began to use personal computers and the attendant software in the 1980s, they needed manuals to understand how to make the machines perform work tasks (e.g., creating, formatting, and printing a document). These early computer manuals were comprehensive but often at the expense of being usable. They were sometimes several hundred pages long and had a poor reputation for user experience.

Printed manuals persist, especially for machinery (e.g., cars), furniture assembly, and household products. Several studies show the continued dominance of user manuals as a format. For example, the CIDM's 2019 report[29] on trends indicates that manuals are the most common content format (almost 90% of respondents). Although manuals continue to be an essential form of user assistance, nowadays they are more often produced for online delivery, via HTML or PDF, rather than in print format.

Topic-based Formats

The industry trend of **component content management** means that you are likely to produce short-form content (like topics) rather than long-form content (like manuals). Online help topics became a popular and prominent form of user assistance in the 1990s. Instead of having to negotiate a 300-page manual, a reader could access a topic specifically dealing with their query. In the past two decades, topic-based writing has become a common content format for technical communicators. Topics are short pieces of text that respond to a single query. Technical communicators write topics that are combined in different ways to produce multiple artifacts. **Single sourcing** describes the principle of using the same source content in multiple outputs. Content needs to be written only once, but it can be reused multiple times. Content is stored in a single-source location (e.g., a CMS). When updates are needed, only the source content is updated, ensuring that the changes are consistent in all outputs.

> You may write content components in a particular genre (like instructional text) that are reproduced in other artifacts. For example, the same content can be reused in a knowledge base article, a PDF, and a website.

This practice reduces redundancy and inconsistencies and should also make translation workflows more efficient because only the source content needs to be translated, so content is translated only once for multiple outputs.

The tools and techniques to support topic-based writing may vary from organization to organization and even within organizations. Writers may write topics using XML and DITA, and often part of the job includes writing metadata, data about the content.

> The reader does not see metadata, but this information helps the content to be selected from a database based on user searches. Metadata might include information about the subject, audience, organization, or author, for example.

Embedded User Assistance

Although online help was the most common form of user assistance for over a decade, users often considered help to be disruptive or even unhelpful. In many applications, help is now *embedded*. This means that the interface includes tips, labels, messages, and other elements that users read in context when they are working on a task. Writers who create embedded assistance work closely with developers to produce intuitive interfaces that have usable and audience-focused labels and messages.

Training and E-Learning Content

Developing training content is a logical extension of the technical communicator's competencies and skills, and many technical communicators develop training artifacts, such as e-learning courses. You may also work as an instructional designer or an e-learning developer (as discussed in Chapter 1). Because you are adept at explaining how to carry out procedures, you can transfer your knowledge to the development of training materials, whether these are online (like e-learning courses) or in face-to-face training environments. Training materials follow instructional design principles and have an instructional sequence, such as Gagné's nine events of instruction.[50] As new instructional techniques like gamification become more popular, some instructional designers create instruction in the form of games. Many instructional designers also work in education, helping teachers and instructors to develop online learning materials.

Websites

Much of the content you develop will be published on websites. Various texts explore website design, usability, and accessibility.[20,24,51] Guidance from these texts includes the following:

- Reduce the number of words. Avoid reproducing long documents online unless you expect readers to print them.
- Use lists and plenty of blank space.
- Use visuals and other multimedia to replace text where appropriate.
- Consider how to ensure that the content is accessible for users who may be impaired by their contexts of use or by disabilities. For example, provide descriptions of images for users who cannot see them. Provide transcripts for users who may not be able to hear or listen to audio.
- Consider the needs of users from cultures other than your own.
- Ensure that the content is responsive since people access web content on myriad devices, browsers, and platforms.

Websites have subgenres, each with their own characteristics.

Video and Animation

Various types of video and animation content are becoming more popular in user assistance. The past decade has seen a surge in user-generated video instructions. Individuals can easily create video using apps on their smartphones, and they can share them using platforms like YouTube. Whereas video was expensive and time-consuming to produce, many tools and apps are now free, and high-quality video is much less costly. One practice narrative described the range of tools they use to create this type of content:

> [I] design and create various multimedia including animations, audio, video, and infographics using Premier Pro, Photoshop, Yvond, Articulate Storyline, Audacity, and Visme.

You need to consider the instructional context when selecting whether to include video, a screencast, or an animation, as illustrated in Box 4.1.

Box 4.1   When to use video, screencasts, or animation

Video content is effective for showing users how to undertake short procedures. Many individuals turn to YouTube first to learn how to do a short manageable task, whether for a hobby or for work.

Screencasts, recordings of on-screen action with a voiceover, have become a popular format for instructional content. They are very effective for demonstrating how to carry out a procedure using software.

Animation is used extensively in instructional content to show procedures or sequences that cannot be shown physically (e.g., biological systems or geographical events). Adobe's popular animation tool, Flash, has been replaced by several alternatives that produce HTML5 content.

For long procedures, text content may be more suitable than video. One study[52] found that users have difficulty searching video content and report that it is easier to read text instructions while undertaking a series of steps. In a long procedure, users may forget the steps after watching a video, and they find text, or text-and-image, instructions easier to follow.

Social Media Content

In many roles but especially marketing communication, you may be responsible for producing social media content about your product or your content. You may publish posts on microblogging sites like Twitter, or you may need to write longer-form blog posts. This type of content is sometimes promotional. Nevertheless, purely promotional content is not popular with readers, who want to be able to use content or learn from it. Many influential bloggers offer their readers added value. For example, they may focus on writing specific types of documentation (e.g., Tom Johnson's *I'd Rather Be Writing* blog), offering tips or information about specific approaches like content strategy (Scriptorium's blog), or offering tips for using specific tools (e.g., *ffeathers*, Sarah Maddox's blog).

If your role involves posting social media content, it is a good idea to evaluate what others do well and what types of social media content are most popular.

Social media content is publicly available and searchable. You should not post any content that you have not verified.

> Even though this type of content is less formal than other formats, it is essential to post accurate and correct information to maintain your and your organization's credibility.

Podcasts

Podcasts have become a popular format for some types of content. Like video content, podcasts are less easily searchable than text. You might use them to provide background content or for additional information about a product. The many freely available recording tools and mobile apps have made podcasting viable for content development projects. Short podcasts with less detail may be more useful than very long recordings. Podcasts also offer scope for creativity. You can create hybrid formats that combine elements of storytelling with more instructional content.

These are just some of multiple possible artifacts that you will develop in your career. Of course, like all aspects of a technical communicator's job, genres and artifacts have evolved as technology has progressed. You can expect continued change in the future as technology evolves. In Chapter 6, you will read about activities, tools, genres, and artifacts that may become prominent in the future.

Summary of Chapter 4

In technical communication, you use a range of competencies and skills, including communication, writing, domain knowledge, and technological literacy. You also need several professional aptitudes, like problem-solving skills, creativity, and detail orientation.

- In the workplace, you will transform your skills to activities. The most common activities are writing, research, information and visual design, management, and application of domain knowledge.
- Although the vignette suggests a possible "typical" day, respondents to the practice narratives agreed that there is no typical day. There was enormous variety in the activities respondents described.
- The range of technical communication tools is expansive, and right now there is no industry-standard content development tool. You will need to be prepared to learn to use a range of tools in industry. Although tools knowledge is important, it is more important to be flexible and willing to learn.
- In some roles, technical communicators need to learn to read or write code (or both).
- You will develop several types of artifact and use a range of genres in any technical communication role.
- Common genres include procedural and report writing and presentations.
- Common artifacts include user guides, topics, embedded assistance, and video content.

Discussion Questions

1. Consider the list of competencies and skills in Table 4.1. Which ones have you already developed? To what extent? Which ones do you need to work on?
2. Re-read the vignette.
   a. List the tools that Oscar uses and identify the purpose of each tool.
   b. Consider the array of activities that Oscar is involved in. Are you surprised by any of these activities? Which ones are most interesting for you?
   c. What competencies and skills does Oscar need to do his job? Consider both the competencies and skills he mentions explicitly and any skills implied by the activities.
3. Consider the tools and technology skills named in the vignette and in the "Tools and Technical Skills" section. Which of these tools have you used? Which ones would you like to learn to use?
4. Find examples of texts from two different genres (e.g., a business letter and a quick start guide). Analyze how the genres differ in their use of language, visuals, and information design.
5. List the different genres you have used in your work or study program.
6. List all the types of artifact you have developed in your work/program to date.

References

1 Hui, A. (2017). Variation and the intersection of practices. In A. Hui, T. Schatzki & E. Shove (Eds.), *The nexus of practices: Connections, constellations, practitioners* (pp. 52–67). Routledge.
2 Henze, B. (2013). What do technical communicators need to know about genre? In J. Johnson-Eilola & S. A. Selber (Eds.), *Solving problems in technical communication* (pp. 337–361). University of Chicago Press.
3 TCBOK. (2019). *Career paths*. https://www.tcbok.org/careers/career-paths/
4 Tekom. (2020). *Competence framework for technical communication*. http://competencies.technical-communication.org/home.html
5 TecCOMFrame. (2018). *The joint academic competence framework*. https://www.tec-com-frame.eu/competence-framework/overview/
6 Institute of Scientific and Technical Communicators (ISTC) (2020). *Careers in technical communication*. https://www.istc.org.uk/professional-development-and-recognition/careers-in-technical-communication/
7 Rainey, K. T., Turner, R. K., & Dayton, D. (2005). Do curricula correspond to managerial expectations? Core competencies for technical communicators. *Technical Communication, 52*(3), 323–352.
8 Cargile Cook, K., Cook, E., Minson, B., & Wilson, S. (2013). How can technical communicators develop as both students and professionals? In J. Johnson-Eilola & S. A. Selber (Eds.), *Solving problems in technical communication* (pp. 98–120). University of Chicago Press.
9 Brumberger, E., & Lauer, C. (2015). The evolution of technical communication: An analysis of industry job postings. *Technical Communication, 62*(4), 224–243.
10 Kimball, M. A. (2015). Training and education: Technical communication managers speak out. *Technical Communication, 62*(2), 135–145.

11 Stanton, R. (2017). Do technical/professional writing (TPW) programs offer what students need for their start in the workplace? A comparison of requirements in program curricula and job ads in industry. *Technical Communication, 64*(3), 223–236.

12 Lanier, C. (2018). Toward understanding important workplace issues for technical communicators. *Technical Communication, 65*(1), 66–84.

13 Hayhoe, G. F. (2000). What do technical communicators need to know? *Technical Communication, 47*(2), 151–153.

14 Bridgeford, T. (2018). Introduction. In T. Bridgeford (Ed.), *Teaching technical and professional communication: A practicum in a book* (pp. 3–17). University Press of Colorado.

15 Johnson-Eilola, J. and Selber, S. (2013). *Solving problems in technical communication.* University of Chicago Press.

16 St. Amant, K. (2013). What do technical communicators need to know about international environments? In J. Johnson-Eilola & S. A. Selber (Eds.), *Solving problems in technical communication* (pp. 479–496). University of Chicago Press.

17 Hoft, N. (1995). *International technical communication.* Wiley.

18 Hofstede Insights (2020). *National culture.* https://hi.hofstede-insights.com/national-culture

19 Gentle, A. (2012). *Conversation and community: The social web for documentation.* XML Press.

20 Redish, J. (2012). *Letting go of the words: Writing web content that works* (2nd ed.). Morgan Kaufmann.

21 Schriver, K. A. (2017). Plain language in the U.S. gains momentum: 1940–2015. *IEEE Transactions on Professional Communication, 60*(4), 343–383. https://doi.org/10.1109/TPC.2017.2765118

22 ASD (2020). *The ASD-STE100 ASD specification.* http://www.asd-ste100.org/about.html

23 Horn, R. E. (1985). Results with structured writing using the information mapping writing service standards. In T. M. Duffy & R. Waller (Eds.), *Designing usable texts* (pp. 179–212). Academic Press.

24 Kimball, M. A., & Hawkins, A. R. (2008). *Document design: A guide for technical communicators.* Bedford/St. Martin's.

25 Schriver, K. (1997). *Dynamics in document design: Creating texts for readers.* Wiley.

26 Friis-Dam, R., & Siang Teo, Y. (2020, April). *What is design thinking and why is it so popular?* Interaction Design Foundation. https://www.interaction-design.org/literature/article/what-is-design-thinking-and-why-is-it-so-popular

27 Pope-Ruark. R. (2015). Introducing Agile project management strategies in technical and professional communication courses. *Journal of Business and Technical Communication, 29*(1), 112–133. https://doi.org/10.1177/1050651914548456

28 Hackos, J. A. T. (1994). *Managing your documentation projects.* Wiley.

29 Center for Information-Development Management. (2019). *Research studies.* https://www.infomanagementcenter.com/resources/research-studies/

30 Kunz, L. (2020, May 29). 40 years in the making. *Leading Technical Communication.* https://larrykunz.wordpress.com/2019/05/29/40-years-in-the-making/

31 Breuch, L. A. K. (2018). *Involving the audience: A rhetorical perspective on using social media to improve websites.* Routledge.

32 O'Keefe, S. (2019, July). The Scriptorium approach to content strategy. *Scriptorium.* https://www.scriptorium.com/2019/07/the-scriptorium-approach-to-content-strategy/

33 European Union, European Parliament and Council. (2006). *Directive 2006/42/EC of the European Parliament and of the Council of 17 May 2006 on machinery, and amending Directive 95/16/EC (recast).* https://eur-lex.europa.eu/legal-content/EN/TXT/?uri=celex%3A32006L0042

34 Johnson, T. (2017, October 12). Question: Which software tools should I use if my goal is to write API docs? *I'd Rather Be Writing.* https://idratherbewriting.com/2017/10/12/which-tool-for-developer-documentation/

35 Moore, L. E., & Earnshaw, Y. How to better prepare technical communication students in an evolving field: Perspectives from academic program directors, practitioners, and recent graduates. *Technical Communication, 67*(1), 63–82.

36 Carliner, S., and Chen. Y. (2018, December). What technical communicators do. *Intercom, 65*(8), 13–16.

37 Write the Docs (2019). *Welcome to our community.* https://www.writethedocs.org/

38 Docs as Code. (2020). *Documentation as code.* https://docs-as-co.de

39 Johnson, T. (2019). The job market for API technical writers. *I'd Rather Be Writing.* Retrieved from https://idratherbewriting.com/learnapidoc/jobapis_overview.html#why-employers-look-for-candidates-who-can-read-programming-languages

40 Carliner, S. (2010). Computers and technical communication in the 21ˢᵗ century. In R. Spilka (Ed.), *Digital literacy for technical communication: 21ˢᵗ century theory and practice* (pp. 21–50). Routledge.

41 Swales, J. (1990). *Genre analysis: English in academic and research settings.* Cambridge University Press.

42 Luzón, M. J. (2005). Genre analysis in technical communication. *IEEE Transactions on Professional Communication, 48*(3), 285–295. https://doi.org/10.1109/TPC.2005.853937

43 Tufte, E. (2003). *The cognitive style of PowerPoint.* Graphics Press.

44 Miller, C. R. (1994). Rhetorical community: The cultural basis of genre. In A. Freedman & P. Medway (Eds.), *Genre and the new rhetoric: Critical perspectives on literacy and education* (pp. 57–66). Taylor & Francis.

45 Killingsworth, M. J., & Gilbertson, M. K. (1992). *Signs, genres, and communities in technical communication.* Baywood Publishing Company, Inc.

46 OASIS. (2018, June 19). *Darwin Information Typing Architecture (DITA) Version 1.3 Part 3: All-inclusive edition.* http://docs.oasis-open.org/dita/dita/v1.3/dita-v1.3-part3-all-inclusive.html

47 Henze, B., Miller, C. R., & Carradini, S. (2016). *Technical communication.* Genre Across Borders. https://genreacrossborders.org/research/technical-communication

48 Selfe, R. J. & Selfe, C. L. (2013). What are the boundaries, artifacts, and identities of technical communication? In J. Johnson-Eilola & S. A. Selber (Eds.), *Solving problems in technical communication* (pp. 19–49). University of Chicago Press.

49 Brockman, R. J. (1998). *From millwrights to shipwrights to the 21ˢᵗ century: Explorations in a history of technical communication in the United States.* Hampton Press.

50 Gagné, R. M., Wager, W., Keller, J., & Golas, K. (1985). *Principles of instructional design* (5ᵗʰ ed.). Wadsworth Publishing.

51 World Wide Web Consortium. (2016). *Web design and applications.* Https://www.w3.org/standards/webdesign/

52 Poe, A. K. (2013). The usability of print and online video instructions. *Technical Communication Quarterly, 22*(3), pp. 237–259. https://doi.org/10.1080/10572252.2013.775628

5   Technical Communication Workplaces

Introduction

This chapter sets out to answer questions you may have about what it will be like to work in technical communication.

- Where will you work?
- How will you work?
- What is your professional outlook?

Three vignettes in this chapter describe technical communication in different sites of practice:

- A contract technical communicator who works from home.
- An individual working within a team of technical communicators.
- A lone technical communicator.

These vignettes suggest potential physical and collaborative environments and workplace cultures. Like the vignettes in the other chapters, they are based on responses from the practice narratives. They are, of course, just three of many possibilities.

This chapter begins with a discussion of research on **workplace studies** in technical communication. You will discover how work in this profession is influenced by **workplace cultures and industry sectors**.

In the second section, you will examine **global and globalized features of contemporary professions**, identify how they impact technical communication, and explore the challenges and opportunities they pose.

In the conclusion of the chapter, a discussion of the **professional outlook** for technical communication explores job satisfaction and labor market trends.

Research into workplaces in technical communication is limited, and this research can also date quickly. Therefore, in addition to discussing academic studies, the information in this chapter emphasizes content from websites, trade magazines, blogs, and other online sources. This chapter enables you to recognize the diversity of technical communication workplaces, to explore different workplace trends, and to visualize your potential future career.

- If you are a **technical communication student**, this chapter will help you to envisage potential workplaces and workplace concerns. It will also help you to recognize how your study program is preparing you for the workplace.
- If you **work in industry as a technical communicator**, you will be able to compare your workplace experiences with those described in the vignettes and to recognize commonalities and differences.
- If you are **considering a career change into technical communication**, you will be able to prepare for the transition. Through reading about others' experiences, you may recognize ways to position yourself for work opportunities in technical communication. You will also be able to determine your workplace preferences.
- If you are a **researcher**, you will be able to use the contents of the chapter and the vignettes as a record of prevailing technical communication workplace cultures.

Workplace Cultures, Sectors, and Environments

What will your workplace and work environment look like if you work in technical communication? Of course, there is no universal answer to that question. Several resources do provide information about workplaces, however, that help us to understand the cultures, sectors, and environments where we work and to research workplaces we may want to join in the future.

Workplace Cultures in Technical Communication

Workplace cultures are the features of the work environment that influence our actions and interactions as well as our perceptions of how we fit into an organization. The culture is shaped by explicit factors, like the industry sector, organizational structures, and expressions of the culture in published plans and strategies (e.g., commitments to inclusivity and accessibility). Sometimes, the physical environment gives us clues about the workplace culture. For example, universities often use artwork or public sculptures, like the plaque in Figure 5.1 at the University of Limerick, to articulate the importance of learning in the organizational culture. The University of Limerick's motto, displayed on this stone plaque on the campus, is "eagna chun gnimh," which translates into English as "wisdom for action." The use of the Irish language in the motto can also be taken as an implicit signifier of the value of linguistic and cultural heritage in this organization.

Figure 5.1 A plaque at the University of Limerick expresses the institution's commitment to learning through a motto on the crest (Photo:Yvonne Cleary)

Workplace cultures also have less obvious but perhaps more impactful features, such as management and leadership styles and unwritten behavior norms. Sometimes, these explicit and implicit features are in conflict. For example, an organization may commit to social justice and fairness in its public documents but simultaneously adopt practices that are unfair or that exclude some individuals, stakeholders, or groups.

How can you learn about the workplace culture of an organization? One of your first questions when starting in a new workplace or undertaking your first job is likely to be "How can I fit in?" Jim Henry (p. 76)[1] explains that:

> We are constantly figuring out how to fit into our workplace in order to accomplish the organization's goals.We daily process hundreds of little clues as to how things are done here, often at the nearly subliminal level.

He recommends learning about a new organization by:

- Observing your co-workers to determine how they work and interact, what tools they use, and so on.
- Taking notes (e.g., at meetings).
- Analyzing the artifacts (e.g., documents and emails) that your colleagues produce.
- Interviewing members of the organization and writing up your observations.

If you have taken a technical communication course or program, you may have conducted an interview with a professional as an assignment to help you to learn more about the profession, the interviewee, and their workplace.That information gives you an insight into the workplace culture.

You should not only seek to "fit in" to your organization, however. Your goal when you enter a new workplace should be to contribute in meaningful ways and ultimately to shape the organization.[1]

Industry Sectors in Technical Communication

Technical communication is an expansive profession, and you will find job opportunities in almost every occupational sector. A TechWhirl blog post in 2018[2] identified these sectors:

- Computer-related (e.g., software, hardware, web, social media, and cybersecurity).
- Science, engineering, or health care.
- Government.
- Instructional design.
- Business or finance.

Most job opportunities are in private industry, where we work in:

- The information technology (IT) sector (e.g., hardware, software, information and communications technologies (ICTs), apps, web services, and social media).
- Financial services and banking.
- Health-care, medical, and pharmaceutical industries.
- Nuclear, aeronautical, and aerospace industries.
- Media and cultural industries.
- A range of other sectors, including hospitality, shipping, textiles, transport, and utilities.

Although most technical communicators work in private industry, public sector roles include, for example:

- Developing e-government content, such as website content that enables citizens to access public services online.
- Developing training and educational content for online course delivery at all educational levels.
- Creating instructions and training materials for use in the public and military services.
- Creating health-care information for use in public hospitals and medical facilities.
- Writing and explaining policy. In the US, for example, "the government hires a lot of tech writers in odd places, and one of the career paths for a tech writer leads to policy analysts, media relations specialists, and many other jobs that require someone to explain complex concepts."[2]

A 2018 *Intercom* census of technical communicators provides insight into work sectors.[3] The census had 676 respondents, 80% of whom were based in the US. Although respondents were employed across all sectors, the majority were in private sector roles. About 20% worked in technical communication consultancy or contracting. Some respondents worked in manufacturing, some in business, and some in medical and health technology, but the majority worked in technology and IT. Brumberger and Lauer[4] also found from their job advertisements research that, although technical communication jobs are advertised in almost every industry sector, most roles are in IT and business services. In government and the public and nonprofit sectors, technical communicators work in universities and education more generally and in nonprofit organizations. In an *Intercom* article, Allen Brown (p. 28)[5] explained how he uses his technical communication skills as a manager in a nonprofit organization:

> I look for any chance I can to apply what I've learned from my technical communication studies. I design, write, and edit job descriptions, policies, procedures, job aids, reports, decision trees, flow charts, communiqués, and marketing collateral, including content for blogs, social media channels, and our website.

Industry Sectors Internationally

Regional and country variations are evident in technical communication, and these variations are driven by the prominence of industrial sectors in different regions and countries. As Ding argues: "we should acknowledge the need to reconceptualize technical communication to meet local needs and to remain open minded for alternative local models and different practices" (p. 302).[6] For example, in many European countries[7] and in China,[8] there are close links between translation and technical communication, regardless of the industry sector. These links have driven how technical communication has developed as a profession in many countries.

Throughout Europe, most technical communication jobs seem to be advertised in computer-related (hardware, software, services, and applications) roles. According to tekom Europe,[9] 90% of (about 90,000) technical communicators in Germany work in software or manufacturing and 10% in technical communication consultancy companies or as contractors. Research that I conducted with practitioners in Ireland[10] found that most respondents (almost 80%) worked in private industry, primarily in IT and business services. Because Ireland does not have a strong history of manufacturing, this was not a prominent sector. In that study, three types of workplace were most prevalent in private industry:

- Large teams of technical communicators in multinational IT and enterprise software companies.
- Writers contracting for dedicated technical communication consultancy firms.
- Lone technical writers or very small writing teams in manufacturing and ICT firms.

Table 5.1 Regions and work sectors

Region	Sectors
North America	The US Bureau of Labor Statistics offers various types of analysis for the occupation of "technical writer," including strongest employment sectors (computer systems design and related services) and states with most job opportunities (California and Texas) and best remuneration (California and Washington, D.C.).[11]
	An analysis of job advertisements showed that, in the US, technical writers/editors were employed across all sectors. The biggest proportion (26%) was in IT services and software.[4]
	The same research found that content developers/managers also had most opportunities in this sector, although this job title was also associated with marketing and social media roles and web and internet services.
Europe	In Germany, technical communicators work in many sectors but are most active in engineering, automotive, consumer goods, software, and medical technology.[9] In many other European countries, technical communicators are most like to work in the IT sector, especially in software and internet services.[12]
	In countries that have a tradition of engineering and manufacturing (e.g., Germany, the UK, and France), technical communication is strong in these sectors.
	In the UK, there is some regional variation. Although some technical communicators based in the north of England develop content for machinery and engineering contexts, in London, Cambridge, and the south of England, most roles are offered in software and web services.[13]
Asia	In China, until the 1990s, technical communication was not a specialized profession, but technical publications were produced and translated for the manufacturing industry.[6] In 2016, about 3,000 technical communicators worked in China, one third of them for Huawei.[8]
	Technical communicators in India have opportunities in the IT sector.[14]
	In Japan, technical communicators have opportunities writing content for consumer electronics and IT applications.[15]

Table 5.1 shows a snapshot of some of the global variety for regions where I could access information.

Learning about Work Sectors

There are many opportunities to learn about occupational sectors. If you have studied technical communication,

- Your instructors may have organized for professionals to present a session about their work in one of your courses.
- You may undertake a technical communication internship.

- You may have completed a service assignment, such as creating content for an organization.
- You may have interviewed a professional for one of your courses.

All of these opportunities to interact with the profession give you insight into workplaces and work sectors. Even if you have not studied technical communication formally, you can learn about work sectors through research studies, blogs, and social media and from job advertisement websites. Workplace narratives also help you to visualize what it would be like to work in this profession. Because technical communication has a limited public profile, the vignettes that feature later in this chapter, along with other workplace stories, help you to imagine what your future work and your career might entail. They can also help you to empathize with other professionals, establish your professional identity, and explain your work and workplace to others.

Work Environments in Technical Communication

Although it is not central to the work you do, knowing something about the physical environment also helps you to envisage your future as a technical communicator. Of course, every workplace is different, and the physical environment depends on many factors, including practicalities such as the industry sector and the organizational culture.

Some corporate videos suggest what these work environments may be like. A tekom Europe video, made in collaboration with SAP Germany,[16] shows a technical communication workplace in Germany. The video begins outside an SAP building, a fairly typical office block. Interior scenes show individuals using laptops in open public spaces, including one shot with a bicycle hanging from the ceiling, suggesting a modern, inclusive, and adaptable workplace. Other scenes are in meeting rooms with projectors and whiteboards, and a couple of scenes show individuals at their desks with desktop computers and large monitors. The overall impression is of a cool and modern workplace. A corporate video from Google[17] has similar workplace scenes, showing technical communicators working in a typical open-plan office environment but also in funkier public spaces and meeting rooms. A man travels through the inside of the building on a scooter, suggesting a relaxed and casual work environment.

> Although these videos depict workplaces in two different countries, the physical environments have a lot in common because of the corporate, multinational workplace cultures in both organizations.

Brumberger and Lauer[18] shadowed nine technical communicators at work. The authors found that workplaces varied vastly, from lone writers in cubicles that would have been typical in corporations in the 1990s to more informal or innovative environments such as those depicted in the corporate videos.

> The research on the sectors and environments within which technical communicators work, and the ways those environments shape their work, is limited. This is an area where more research is needed to help students and new practitioners to prepare for their future careers.

Vignette 1: The Content Writing Team Member

The first vignette in this chapter describes a technical communicator who works in a large corporate Content Writing team and in a corporate work environment similar to those depicted in the Google and tekom Europe videos. This vignette is based on composite descriptions of similar workplaces in the practice narratives.

> *I work as a Content Writer for a well-known multinational software company. About 30 of us work as Content Writers at our site, so we are a team, but as individual writers, we are also embedded in other development teams. We are also part of a larger team of 200 Content Writers based in other locations around the world. My workspace is a dedicated desk in a large open-plan room that houses our entire Content Writing team. I have a desktop computer with a large monitor and a laptop that I use as a second monitor. The large screen is necessary for close reading and for looking at screenshots and test environments.*
>
> *Our building is a fairly typical example of corporate real estate. It's over five floors, with several open-plan public spaces that have comfortable seating and good Wi-Fi, so you can actually work anywhere in the building. Sometimes, if it's getting noisy in the office or if I just want a change of scene, I'll sit in one of the building's open spaces and work on the laptop.*
>
> *I clock in and out every day, and my manager always knows when I am in the office and how long my breaks take since her office is just off our open-plan area. There's reasonable flexibility in booking holidays, and I have 25 days holiday per year, plus eight public holidays, a perk of working in Europe.*
>
> *The development team is based in another country, and I am the only Content Writer on that team. I dial in for scrum meetings daily. These meetings help me to stay on top of changes that will impact my work, and they also help me to be involved and to represent product users. Every other Wednesday, the site Content Writing team has a meeting, but because the whole team is in one office, we get to share information, find out what's going on, and ask each other questions regularly, not only in those meetings. This is a great resource, especially as this is my first job and I have a lot of questions. We also have occasional meetings of the Global Content Writing team. Because I have a good internal network, I don't feel the need to join a professional association or to network with other technical*

> communicators. *I do know one technical writer outside the organization whom I met through a book club.*
>
> *I'm not sure how my career will progress here. If I want to move into management, there's just one manager for 30 people. After seven years, I will become a Senior Content Writer, but I will essentially be doing the same job, just with a higher salary. I am not planning to stay in this job for that long, but this is my first technical communication job, and I'm glad I started in a supportive environment that has helped me to learn and given me access to a network.*

This vignette shows that an open-plan space can facilitate engagement and creativity for technical communicators. A TechWhirl blog post[19] described other advantages of an open-plan workplace.

> I hear a lot of complaints about open-plan, but it works for me. I can keep tabs on key people much like sitting next to a meeting room, but with line-of-sight, I'm more proactive, … scanning for high-value targets. I spend more time getting answers and writing doc and less on waiting for responses via email or Slack.

Despite the trend toward corporate workspaces, many of us now work from home and design our own working environments. Because of affordances of technology and infrastructure as well as the Covid-19 pandemic, working from home is increasingly common. This is discussed in more detail in the next section.

Global and Globalized Features of Contemporary Work

Research about contemporary professions is abundant. Several themes from that research have resonance for technical communicators. In this section, we explore trends and themes in the contemporary workplace and their impact for technical communication: remote working, the related increase in virtual teamwork, outsourcing, offshoring, and restructuring. They have the common thread of being facilitated and driven by technology and globalization. Understanding these trends helps you to recognize how you are preparing for globalized workplaces and to determine what additional training you might need.

Remote Working

Remote working refers to the practice of working at home or in another location for part or all of your contracted work time with the support of technology. Lanier surveyed technical communicators to identify the most important concerns of practitioners, one of which was remote working. He observed that "technical writers can perform their work from any location, and so associated trends include rising instances of remote working or of organizations hiring contractors and third parties that work at a distance" (p. 81).[20]

Remote working is increasingly facilitated by developments in technology and broadband/wireless infrastructure. As early as 2005, Jablonski noted how "traditional boundaries, like those dividing workplace from home and recreation, are being permeated" (p. 15),[21] as technology enabled individuals to work from home and, by extension, to spend more of their leisure time on work tasks. Jablonski's description of the bleeding of boundaries has become increasingly relevant, as so many corporate structures are globalized and many employees do some forms of work from home.

> During the Covid-19 pandemic, so-called "lockdowns" and "stay-at-home" or "shelter-in-place" orders worldwide forced many millions of workers to switch to short- to medium-term home working.

On a single day in April 2020, Microsoft logged 4.1 billion minutes of meeting time on Teams, the Microsoft remote meeting application.[22] Microsoft estimated that the mass movement toward home working forced by Covid-19 led to "two years of digital transformation in two months." Their data suggest that the trend will not reverse entirely after the pandemic has passed.

> Although remote working is not possible for all employees and is not facilitated in some workplaces, as a technical communicator you may have opportunities to work from home.

Remote working has several advantages, as described in recent research and articles on the subject[23,24,25,26] and in the practice narratives. These advantages include the following:

- **Improved performance**, possibly due to fewer distractions, increased independence, and an awareness of being trusted. One practice narrative explained how their home working time was more productive than time in the office:

  I work from home on Tuesdays and Fridays; those are the days that I do by far the most writing.

- **Time and environmental savings** through not commuting to work. For another practice narrative respondent, a 90-minute commute to the office was a compelling reason to work from home two days per week. The European Commission also identified environmental benefits as a positive outcome of remote working.[26]
- **Reduced traffic** and congestion on public transport, accruing benefits for those who must travel to their workplace.

- **Flexibility**, which may be important for employees at different life stages. A practice narrative explained:

  > I pretty much set my own hours around the occasional meeting. My company allows me to be really flexible, which was the reason I took this job, so I could care for my elderly mother, as well as other personal/business endeavors on the side.

- More **buoyant rural communities** since people do not have to live in urban areas to access work opportunities. In a blog post,[27] Tom Johnson explained why this would be an advantage for him:

  > And if people can work anywhere, we'd transition into a globally distributed workforce. I'd definitely move … into a much more cost-friendly, livable city, to a place where I can buy a house and maybe ride a motorcycle around.

- **Increased job satisfaction**, although the research shows that this issue is complex and depends on many factors; remote working does not always increase, and may sometimes decrease, job satisfaction. The "Job Satisfaction in Technical Communication" section below explores the complexities in more detail.
- Improved perception of **work–life balance**, although, again, the converse was also reported.
- **Lower stress**, which was associated with a reduced sense of time pressure. Conversely, many individuals, especially parents and those with caring responsibilities, reported higher stress when they switched to home working during the pandemic.

The same sources examined the challenges posed by working from home, including during the Covid-19 pandemic. They reported challenges such as:

- Complexity of **monitoring and recording performance**. Managers need to find new ways to monitor productivity rather than simply focusing on presence or absence and how that is recorded. Research suggests that it may be better to measure work outputs than to attempt to monitor time spent online. Managers may need training in managing employees who are working from home. In the same vein, employees may need to keep more thorough records of how they spend their time. In Europe, there is concern that monitoring individuals working in their homes has unacceptable privacy implications.[26]
- **Isolation**, both social and professional. As humans, we crave interaction, and often a face-to-face encounter is more productive and meaningful than online collaboration.
- Likewise, there may be an adverse **impact on teamwork** and relationships with co-workers. As with other benefits and challenges, this impact has both positive and negative associations. Some research suggests that less interaction

may in fact lead to better relationships with co-workers in some cases. Research on virtual teams (discussed in the next section) shows that it is possible to set up effective teams where members do not meet face-to-face.

- Remote workers report concerns about how a **lack of visibility** impacts their career opportunities and progression. Remote working may not suit early-career employees who are just beginning to develop a profile and a support network and community.
- Remote workers need **infrastructure** in order to be able to work from home. In some cases, the costs associated with working (electricity, heat or air conditioning, and internet connection) are borne by employees instead of by the organization.

Several variables influence the success of remote working: the type of work, management styles, and the characteristics of the individual worker.[23] For example, remote working might or might not suit your personality. Flexible communication patterns and increased training could help to make remote working and virtual teamwork successful.

Vignette 2: The Contractor who Works from Home

Although I collected the narrative data in late 2019 (before the impact of Covid-19), many practice narrative respondents were already working from home at least some of the time. In this vignette, a contractor who works from home describes their situation; again, this vignette is based on composite responses. In a vignette later in the chapter, a lone writer works from home a couple of days each week, a common scenario in the narratives of lone writers.

I studied Computer Science and worked in software engineering for a decade. I had some interaction with technical writers and occasionally had some writing assignments in that role that I really enjoyed. I took a career break to raise my family, and when I was ready to return to work, I decided to make a more formal move into technical communication. My software engineering roles were not family-friendly, whereas I had the impression that technical communication could be. You could work from home or part-time instead of having a full-time job. That decision led me to take an online Graduate Certificate in Technical Communication.

For the past two years, I've had two short-term contracts with software companies, working as the sole technical communicator (I am called a technical writer). I work primarily from home. If I work on site, it tends to be for occasional days. I have a home office, and having a dedicated space to work, where I go in the mornings after breakfast and leave in the evenings when I finish, is really important psychologically. It helps me to separate work from home. When I started working from home, I worked at my kitchen table. My home life distracted me from work, and I would find myself doing laundry or other household chores when I should have been working. I also sometimes worked late into the evenings to finish work, because I was missing a psychological separation of work from home. Now I make sure to structure my day, to take a lunch break (when I can do those chores!), and to finish at about the same time daily.

I have a high-spec laptop, a desktop computer and a couple of monitors, but for my current contract I use a company laptop that has secure access to the network, software, and systems. I rely on my home Wi-Fi for connectivity. I get a tax rebate for costs associated with work, including the Wi-Fi subscription. Almost all my meetings are online, through Microsoft Teams or on the phone, and a lot of other communication is by email and direct messages.

Because I work from home and for myself, I can take time out during the morning and afternoon to drive my children to and from school. I can also fit other personal appointments into my day if I need to. While I appreciate the flexibility of working from home, I miss the sociability of an office. My ideal working week would be four days at home and one day in an office, to keep in touch with people and to have a little more variety in my days. Another disadvantage of working as a contractor is that I get paid only for the days I work. I need to factor holidays and training in when I price contracts, because I will not be paid for that time.

Virtual Teams

Virtual teams are groups of two or more individuals in different locations using technology to work on a shared task. As workplaces become more globalized and members of a work team may be in different buildings, cities, or even countries, virtual teams have become commonplace in organizations. Because remote working has become more prevalent and sometimes necessary, the need for virtual teamwork has also increased.

> In work or study, you may have worked on assignments with teammates you never met, using email, videoconferencing, or other technologies to collaborate: this was virtual teamwork.

Virtual teams have many benefits for organizations. They reduce travel costs while enabling experts to work together even if they are not located in the same city or country. Studies of virtual teams demonstrate that several factors make them more effective, including the following:

- Socioemotional **communication** (e.g., sharing some personal information and having time for some small talk at the start of an interaction, especially in the early stages of the team formation). One study of a student virtual team found that socioemotional communication happens through non-task communication, humor, and even conflict.[28]

- Strong team **leaders**, who recognize the challenges members face in a virtual team and who make explicit efforts to connect with the team.[29]
- **Trust**, which needs to be established quickly because of the short time frame of many projects. Trust usually develops on the basis of scope for communication, individual personality traits, and performance.[29]

Virtual teams are enabled by technology. Respondents to the practice narratives reported collaborating with virtual teammates using multiple technologies. Among those listed were email, instant messaging, project management software, and web conferencing tools like WebEx, Microsoft Teams, Zoom, and GoToMeeting. In any team, members need to agree about which tools to use for which purposes, and technology needs to be reliable for effective teamwork. It is essential to ensure that communication is as clear as possible since communicating through technology is less context-rich than face-to-face communication. You have fewer nonverbal cues to enable you to recognize the tone or intention of a message. These cues are more explicit in face-to-face communication. One study found that "nuanced cues that can be picked up in often-seen and familiar facial expressions, tone of voice, or interaction behaviors of face-to-face employees are unavailable" in virtual teams (p. 31).[29] Additional challenges to virtual teamwork include dealing with time and cultural differences. Several practice narrative respondents described how they work with colleagues in different countries, as in these excerpts:

> We use email, face-to-face interaction and Microsoft Teams for instant messaging. Many of the developers are based in Spain, and some SMEs [subject-matter experts] are based in the UK, so I use email [and] Teams a lot.
>
> I became part of a content team that contained two managers, an editor and four other writers: two at our company base in the north of England; one in London; one, Stateside. Since then, the content team has expanded to encompass an additional writer and another editor, plus our own engineer and two localisation specialists.

Organizational Practices that Impact Technical Communication

Various organizational practices, many a result of the interactions of globalization and technology, will affect your technical communication career.

Offshoring is the practice of moving business units to cheaper countries. It was a consistent trend in technical communication in the 2000s and is enabled by technology and driven by globalization. During that decade, it was a real concern for US-based technical communicators and organizations. Paretti, McNair, and Holloway-Attaway explained that "outsourcing, offshoring, and globalization, enabled by a dynamic network of communication technologies, have altered the physical and social landscapes of our working lives" (pp. 327–328).[30] At the same time, this practice saw the profession flourish in countries like Ireland and India, where technical communication and other services were regularly offshored, at least in part because wages and corporation taxes were relatively low. Consistent policies of offshoring mean that jobs are never secure in any location since firms will ultimately move to cheaper locations when it is expedient. Because technical communication requires strong language skills (most often in English), this profession may be less susceptible to offshoring than some others. Nevertheless, some practice narratives were concerned by the trend, as in this excerpt:

> I'm not entirely optimistic: given what I see in my current company, technical writers are primarily recruited in India. In fact, my current company will only recruit technical writers in India.

The more recent trend of "re-shoring" sees "the movement of offshored production back to its previous location" (p. 22).[31]

Outsourcing refers to the practice of contracting work to an outside organization or supplier instead of having an in-house employee do the work. This trend seems to be more prominent in technical communication during recessions, as Greg Wilson (p. 82)[32] explained:

> When I worked as a technical communicator in the first half of the 1990s, large computer companies in my state were laying off technical communicators along with other employees who did not contribute directly to profits. Small technical communication consulting firms sprouted to contract with these companies, and soon, the only jobs to be had were through these contracting firms.

I interviewed technical communicators in Ireland in 2010, in the depths of a recession, and found a similar trend. One interviewee who worked as a freelance technical communicator at that time told me,

> One job I got [because] the company had pared back their staff. They had one writer. But they had no contingency and there was a product they needed to release, so they needed someone for two months. That was a consequence of the recession. A year or two earlier, someone in the company who was free could have taken up the slack.

This practice continues, in some organizations at least. A practice narrative respondent who worked for a consultancy firm made a similar observation:

> I know from experience that internal [documentation] departments are usually the first to go when money gets tight. Working for a company that ONLY does this means when other companies reduce or eliminate their human resources, they will often look for a less expensive alternative. We are that solution.

In technical communication, service companies have emerged in many countries in response to the trend of outsourcing.

Offshoring interacts with outsourcing because work is often outsourced to external service providers based in other countries (or offshore). Lanier calls it "the business strategy of outsourcing certain business activities (such as technical writing) to international service providers" (p. 78).[20] In spite of this trend, the 2018 *Intercom* census of technical communicators, most of whom were based in the US, found that respondents "feel secure in their jobs and do not seem concerned about the impact of outsourcing" (p. 27).[33]

Corporate restructuring is a common feature of the modern workplace and a function of globalization, mergers, acquisitions, new management, and other internal and external factors. It involves either a shallow or deep reorganization of departments. It can lead to offshoring and outsourcing because if your organization restructures, some departments might close and their work must be outsourced or offshored or both. Restructuring is common not only in corporations but also in education and other sectors.

> One practice narrative respondent explained how they found themselves working for their first employer a decade after leaving, following acquisition of the smaller organization to which they had moved.

Restructuring can have an impact on the visibility of your role, as one practice narrative observed:

> Every time there has been a change in the past, my job gets pushed further and further down the org chart to where I have no authority anymore.

Restructuring can also lead to a great deal of uncertainty, as Tony Bove, an Ottawa-based technical writer, explained in an *Intercom* "My Job" feature (p. 40):[34]

> As soon as I hit my stride, it was announced that [the product line] was to be transferred from Healthcare to the Global Tools and Storage (GTS) business unit. UGH! Uncertainty. I didn't know if the new group already had technical writers, if they even wanted a technical writer, or if my position would be eliminated or moved to the United States, where GTS was headquartered.

He described how further restructuring meant that he was the only member of his business unit based in Canada, requiring him to report to a manager in a different country. Six months later, he was asked to work from home and later still he was transferred to a different unit. You will likely have to deal with some uncertainty due to restructuring at some point in your career. Bove's lesson from his various experiences is "when uncertainty raises its ugly head, stretch" (p. 40).

Additional relevant themes in the research on organizations include the following:

- **Diversification.** This trend affects all professions in the modern era.[35] It has been widely discussed in technical communication, and different aspects of this trend are discussed throughout this book.
- **Workplace learning.** This trend is discussed further in Chapter 2.
- **Automation** and the impact of artificial intelligence and big data. This trend is discussed further in Chapter 6.
- More **casual employment** modes. This trend is also discussed, with respect to its future impact on technical communication, in Chapter 6.

Vignette 3: The Lone Writer

The final vignette in this chapter describes the work of a lone technical writer in an IT services company, who has experienced corporate restructuring. This vignette is a composite of responses from lone writers to the practice narrative questions.

> *I am the only writer among a staff of more than 200 in a small business unit of a larger multinational organization. I go into the office to work about three days per week, and have flexibility to work from home the other two days. In the office, I work in an open-plan cubicle environment, and at home I have a home office and use my work laptop. Although I have worked for the 'same' employer for more than a decade, in that time I have moved department three times, the first time because a new manager wanted to move writing from Development into Marketing. After six months, I was moved back to the Development team. The third change was when our department was merged with another one due to a corporate acquisition. These changes are unsettling, and we spent a lot of energy worrying and projecting, but in reality, the restructuring has not had much impact on my day-to-day work.*
>
> *My job title is Senior Technical Writer. My main role, for which I was hired, is writing and updating operator and maintenance handbooks for our software and hardware. Because I am the only writer in the organization, I do various other writing tasks. People from all parts of the organization contact me when they need some kind of content, mostly text. I write articles for our website and newsletter, produce reports, and help other teams with any writing they need done. I also help with knowledge gathering from the technical and business SMEs. I sometimes meet with company executives about current projects or initiatives and document those projects in whatever way is necessary (internal or external communications). Sometimes, however, individuals have expectations that I have to correct, such as that I am an editing resource for them, or that I should be available on-demand for their work. But because I have now worked here for more than a decade, most people understand what I can and cannot do.*

I choose my own software tools, and I have a fair amount of autonomy over how I work, but I do have to work to meet the deadlines of other teams. The job is busy because of the variety, but as long as I meet deadlines, I do not have to work overtime.

I miss having someone to review my work, and someone to discuss writing decisions with. Those are the main drawbacks of being a lone writer. I use a well-known corporate style guide to be consistent in my writing decisions. To get the perspectives of other technical communicators, I network fairly actively through LinkedIn, and I attend professional conferences. I became involved in the STC's LW SIG (Society for Technical Communication's lone writer special interest group), and this membership helps me to network and to feel part of a community. My manager is on the product development team and is very supportive whenever I request training, and I am funded to attend at least one conference annually.

As a lone writer, I have to work harder to get noticed, to ensure people know that what I do is important, and to ensure that I am not passed over for promotion or other opportunities. For that reason, I have resisted working from home full-time. Advice from members of the LW SIG has helped me to position myself more effectively within the organization.

Professional Outlook for Technical Communication

As you embark on a new career, you probably want to know about the professional outlook. In this section, we explore job satisfaction and labor market trends.

Job Satisfaction in Technical Communication

In the 1990s and 2000s, job satisfaction was a concern in technical communication. Research on professionalization showed that many practitioners felt undervalued. As Henry explained, they were often excluded from information circles, to the detriment of doing their work (p. 81):[36]

> Perhaps because of their frequent secondary status within organizations and certainly because they are often located outside inner circles of information and knowledge in organizations, writers may encounter disparities between the information they receive and the information necessary for efficient and effective document processing.

Career stagnation was also a concern that led to the perception of technical communicators having a lower status than other professionals.[32] Recently, however, as technical communicators have become more embedded in development teams, the profile and status of the role seem to have increased. My research with Irish technical communicators indicated that, in Agile teams, the visibility and involvement of writers within the development team led to increased recognition for the role. A manager of an Information Development team explained why project teams that use an Agile framework can have positive impacts on visibility:

> All the authors are assigned to scrums, the software is developed in sprints, with maybe 10 or 12 sprints to a release. Each scrum will have one [technical communicator] assigned to it. You hear mixed reports but overall it's been very positive. For the first time we are now very much part of the team. It's easier to get yourself on the agenda. You're there. You're planned in. It's all visible. So that's really positive.

Lanier's research[20] agrees that technical communicators are now more involved with product development, particularly in Agile teams, and, as a result, may have a more prominent profile and higher status within teams than in previous decades. (Chapter 4 discusses Agile work in more detail.)

Based on their *Intercom* census data, Carliner and Chen concluded that technical communicators are satisfied with their jobs in the main and perceive their work to be respected by colleagues.[33] The practice narratives also suggest a broad trend toward job satisfaction, for the following reasons:

- **Flexibility** in terms of projects, hours worked, or opportunities to work from home.
- **Variety** in projects, tools, tasks, and skills.
- **Salary** and other benefits, commensurate with the skill required and the quality and value of the work.
- **Autonomy** to choose work and prioritize tasks. Lone writers and contractors, in particular, valued their independence.
- **Contribution** of the work. One respondent explained ways in which they had added value:

> One of my customers hired someone from a major government department who subsequently told us that the bid my team wrote was used as the standard by which all the other bidders were assessed. For that bid and for most of their tenders for the time he was working there.

Owing to the relatively small number of responses (62 in total), the range of countries (13), and types of work undertaken as well as the qualitative approach and the different emphases in different practice narratives, I could not determine the extent to which practitioners in different countries, regions, sectors, or roles were satisfied with their jobs. Although that information would be interesting and useful and could be the subject of future research, job satisfaction depends on many personal factors, too. These include career stage, personality traits, level of expertise, colleagues, management, and workplace culture in the team and organization.

Factors that Cause Dissatisfaction

Of course, some practice narrative respondents expressed negative sentiments about their work. I have summarized the factors that caused dissatisfaction at work, according to this and other sources.

Three practice narrative respondents complained about a **lack of opportunities for professional development**. One observed:

> It's been a point of contention with me that I haven't really had any professional development myself.

Carliner and Chen[33] also noted that professional development opportunities have an impact on job satisfaction. Two practice narrative respondents complained about not having sufficient **time** to complete work to a high standard. This excerpt explains:

> I can do some basic editing but I know it is not documented to a standard that I would be satisfied with … This annoys me a bit! I dislike releasing documentation that I am not happy with :)

As we saw earlier in the chapter, many corporate practices can lead to job insecurity. Two narratives mentioned **job insecurity** as a concern, as in this excerpt:

> In terms of personal challenges, in my experience redundancy is an ever-present threat for anyone working in the software industry – perhaps this is the same for all industries.

Two practice narratives also described **chaotic work environments** that impeded their work and caused dissatisfaction. One example described an extreme situation:

> There are always crises from unexpected quarters that immediately derail whatever plan I had for the day. In part, it's my own fault for years of problem solving and keeping a level head – now problems that could go to someone else come to me. … We all want other jobs, but we can't stand the thought of leaving each other. It's combat stress, PTSD [post-traumatic stress disorder] where the stress is renewed daily, year after year.

As negative as this excerpt is, it reflects some important lessons. It signals that technical communication, like other services, can be taken for granted. As Chapter 4 shows, problem solving is an important aptitude among technical communicators. In small organizations or units or in work environments that do not have clear systems in place, you may be called upon to solve problems in projects and areas outside your expertise or remit. If you are consistently solving other people's problems, however, your own work will suffer. The respondent in the last excerpt believes that, by not leaving, they may be complicit in the situation.

The dissatisfaction expressed in some of these comments indicates the importance of reflecting on your work. Sharing your experiences privately with a colleague or someone in your community or network can help you to gain perspective on your job and to determine whether your environment is reasonable and your job satisfying. Anecdotally, graduates have told me they left the technical communication profession because their first role was problematic, chaotic, unsupportive, or overwhelming. It is important to know that such work environments are not normal; moreover, as we will see in the next section, the labor market in technical communication is strong. You do not have to remain in a job that is causing you stress or dissatisfaction.

Labor Market Trends in Technical Communication

The labor market is strong worldwide and professional associations have made steady progress in creating recognition for technical communication as a profession in spite of its relatively low public profile. Figure 5.2, a screenshot from the US *Occupational Outlook Handbook*,[37] shows figures for job openings, pay rates, and growth outlook. The Bureau of Labor Statistics has included "technical writer" as a Media and Communication profession since 2010.

As Figure 5.2 shows, the summary is positive. The growth outlook is faster than for most professions, the number of jobs is high, median yearly and hourly pay is generous, and most technical writers work full-time.

Job Opportunities for Different Job Titles

As we saw in Chapter 1, we use a range of job titles in this profession. I analyzed recent research,[4,18,38,39] websites of professional associations, and job search engines to try to identify the range of job titles. Contemporary job titles tend to have some combination of the prefixes and roots in Table 5.2. I clustered the prefixes and roots to capture the most common combinations. Some job titles use two prefixes, the first often signifying the domain or level of the role (e.g., "digital content specialist" or "senior information developer"). The titles in this table do not include roles in niche specialisms, like content development for the pharmaceutical or financial services sectors.

I also conducted searches of online recruitment sites to determine the job opportunities for different job titles. Most technical communication roles are offered in private industry, regardless of the region. USAJOBS,[40] a database that advertises government jobs, returned just nine vacancies for a search for the term "technical writer" in September 2020. By contrast, thousands of roles were advertised for the same term on recruitment sites that advertise roles in private industry. I conducted searches on both Monster and Indeed in September 2020. Table 5.3 shows the search results for popular job titles.

OOH HOME | OCCUPATION FINDER | OOH FAQ | OOH GLOSSARY | A-Z INDEX | OOH SITE MAP

Search Handbook [Go]

OCCUPATIONAL OUTLOOK HANDBOOK

Occupational Outlook Handbook > Media and Communication >

Technical Writers

PRINTER-FRIENDLY

| Summary | What They Do | Work Environment | How to Become One | Pay | Job Outlook | State & Area Data | Similar Occupations | More Info |

Summary

Quick Facts: Technical Writers

2019 Median Pay ⓘ	$72,850 per year $35.03 per hour
Typical Entry-Level Education ⓘ	Bachelor's degree
Work Experience in a Related Occupation ⓘ	Less than 5 years
On-the-Job Training ⓘ	Short-term on-the-job training
Number of Jobs, 2019 ⓘ	58,400
Job Outlook, 2019-29 ⓘ	7% (Faster than average)
Employment Change, 2019-29 ⓘ	4,300

Technical writers routinely work with other technology experts.

What Technical Writers Do

Technical writers prepare instruction manuals, how-to guides, journal articles, and other supporting documents to communicate complex and technical information more easily.

Work Environment

Most technical writers work full time. Although technical writers work in a variety of industries, they are concentrated in the computer and management, scientific, and technical industries.

How to Become a Technical Writer

A college degree is usually required for a position as a technical writer. In addition, knowledge of or experience with a technical subject, such as science or engineering, is beneficial.

Figure 5.2: Bureau of Labor Statistics, US Department of Labor, *Occupational Outlook Handbook,* Technical Writers, at https://www.bls.gov/ooh/media-and-communication/technical-writers.htm (accessed November 13, 2020)

Table 5.2 Job titles matrix

Domain/level prefix	Prefix	Root
Digital Front-end Junior/Senior Online Social media Web	Content	Administrator Advisor Analyst Architect Coordinator Curator Editor Designer Developer Manager Producer Specialist Strategist Writer
	Document/Documentation Publications	Coordinator Manager Specialist Writer
	Usability User assistance User interface User experience	Consultant Developer Editor Specialist Strategist Writer
	Information	Architect Designer Developer Editor Engineer Manager Writer Specialist
	Technical Medical/science Marketing content	Author Editor Communicator Trainer Writer
	Instructional E-learning	Coordinator Designer Developer Specialist Trainer

Table 5.3 Jobs advertised by title and recruitment site (September 28, 2020)

Job title	Monster	Indeed
"Technical writer"	1,526	1,909
"Technical communicator"	37	35
"Content curator"	5	36
"Content developer"	154	210
"Content strategist"	160	379
"Content writer"	216	511
"Information architect"	92	166
"Information developer"	8	10

At the time I conducted these searches, many jobs were advertised for remote or temporary remote working (in keeping with trends and health advice during the pandemic). Furthermore, in spite of a global and generalized trend toward more precarious and casual working conditions, most jobs were advertised as full-time/direct-hire contracts.

The table shows that in spite of diversification and movements to rebrand, the label "technical writer" still has most traction in industry, but many newer job titles are also prominent. As observed in a study of job advertisements,[4] some job titles are common in specific occupational areas:

- "Content strategist" jobs included user experience in the title or job description in many cases (e.g., 22 of the Monster advertisements).
- "Content developer" roles tended to be in training development.
- "Content writer" roles tended to be in marketing and social media.

International Trends in the Technical Communication Labor Market

I also conducted some searches on regionally relevant recruitment sites and sites that advertised jobs in prominent sectors. These sites have different filters and search strategies, however, so the summaries that follow are indicative of broad patterns, but they are not scientifically validated.

The European Commission's jobs database, EURES,[41] enables you to search for jobs throughout the European Union (EU). Over 16,000 European employers have registered to use the service. Searches for "technical writer" and "technical communicator" in September 2020 indicated a strong labor market throughout the EU, and most jobs were advertised in Germany, Belgium, and Czechia. This is good news if you are interested in working as a technical communicator in Europe!

The IT Jobs Watch website monitors trends in IT jobs over time in the UK, including details such as salaries and live vacancies. This site uses a ranking system based on the number of vacancies corresponding to a job title, where a rank of 1 is the job title with the most vacancies. Job titles related to technical communication ranked around 650 (of over 7,800 job titles). Figure 5.3 shows the

Figure 5.3 The results of a search for "content" jobs on IT Jobs Watch, at http://itjob-swatch.co.uk(accessed November 13, 2020)

results of a search for "content." Notably, although "content management" had dropped in the ranking in the previous six-month period, it had a much higher ranking than the other "content" or technical communication job titles and the highest number of live vacancies at 74 (e.g., compared with 8 for "content strategy").

Using Naukri[14] to search for "technical writer" jobs in India, I found almost 1,000 advertisements (September 2020), most based in Delhi, Hyderabad, Mumbai, and Bengaluru, the software development hub in India.

These snapshots show that in many regions of the globe, technical communication roles are abundant. Academic research about international workplaces, experiences, and opportunities in technical communication is limited, however. More research would enable international students and practitioners to understand and prepare for local and global technical communication conditions and to plan for their future careers.

Summary of Chapter 5

Technical communicators tend to work in private industry and, as a result, have access to the advantages and challenges of corporate working environments.

- Technical communicators typically work in dynamic workplace cultures.
- Although they are most likely to work in the private sector and in software and business sectors, technical communication roles are offered in almost every economic sector, including government, education, and nonprofit.

Future research could examine how the job profile differs in these different sectors.

- Several global and general workplace trends have an impact on the profession of technical communication. These trends include working in virtual teams, remote working, outsourcing and offshoring, and restructuring.
- Vignettes, based on the practice narratives, suggest that different work profiles and work environments suit individuals at different career stages.
- Job satisfaction in this profession seems high. Many factors influence job satisfaction, including flexibility, variety, and fair compensation and benefits.
- The labor market for technical communication is strong. The longer-term outlook depends, at least in part, on strong local and global economies.

Discussion Questions

1. Consider a workplace you are familiar with (e.g., from a part-time job or a previous or current job). Describe its workplace culture. Use sources such as its website and corporate policies as well as your personal experience to inform your description.
2. Use the sources in this chapter to find examples of technical communication workplaces. What are the features of those workplaces? Consider indicators of the physical environment, workplace culture, and global and globalized features of the work.
3. From your reading of the three vignettes, list the key features of each type of workplace described. What, in your opinion, are the benefits and challenges of each of these work scenarios?
4. Find an example of a technical communication job profile online. Based on your reading of this profile, list the characteristics of the job described. Write a short reflection describing what you would enjoy, and what you would find challenging, about this role.

References

1 Henry, J. (2013). How can technical communicators fit into contemporary organizations? In In J. Johnson-Eilola & S. A. Selber (Eds.), *Solving problems in technical communication* (pp. 75–97). University of Chicago Press.

2 Paz, J. (2018, April 17). Where the jobs are in technical writing. *TechWhirl*. https://techwhirl.com/where-the-jobs-are-in-technical-writing/

3 Carliner, S., & and Chen, Y. (2018, December). Who technical communicators are: A summary of demographics, backgrounds, and employment. *Intercom*, *65*(8), 8–12.

4 Brumberger, E., & Lauer, C. (2015). The evolution of technical communication: An analysis of industry job postings. *Technical Communication*, *62*(4), 224–243.

5 Brown, A. (2018, January). Nonprofits need technical communicators, too. *Intercom*, *64*(1), p. 28.

6 Ding, H. (2010). Technical communication instruction in China: Localized programs and alternative models. *Technical Communication Quarterly*, *19*(3), 300–317. https://doi.org/10.1080/10572252.2010.481528

7 Cleary, Y., Engberg, J., Karreman, J., Meex, B., Closs, S., Drazek, Z., Ghenghea, V., Minacori, P., Müller, J., & Straub, D. (2017). TecCOMFrame: Building bridges between technical communication and translation studies through a prototype specialisation curriculum. *Lebende Sprachen, 62*(2), 313–332. https://doi.org/10.1515/les-2017-0025

8 Ding, H. (2019). Development of technical communication in China: Program building and field convergence. *Technical Communication Quarterly, 28*(3), 223–237. https://doi.org/10.1080/10572252.2018.1551576

9 Tekom Europe (2020). *Deciding on your future.* https://www.technical-communication.org/technical-writing/outline-of-technical-communication/perspectives

10 Cleary, Y. (2016). Community of practice and professionalization perspectives on technical communication in Ireland. *IEEE Transactions on Professional Communication, 59*(2), 126–139.

11 Bureau of Labor Statistics. (2020). *Occupational employment and wages May 2019, 27–3042 Technical writers.* Retrieved November 10, 2020, from https://www.bls.gov/oes/current/oes273042.htm

12 Straub, D. (2019, January). Technical communication across Europe. *TC World.* https://www.tcworld.info/e-magazine/education-and-training/technical-communication-across-europe-952/

13 Cleary, Y., & McCullagh, M. (2020, July). Technical communication in the United Kingdom: The academic and professional contexts. In *Proceedings of the 2020 IEEE International Professional Communication Conference (ProComm).* IEEE. https://doi.org/10.1109/ProComm48883.2020.00007

14 Naukri. (2020). *Find a job at India's No. 1 job site.* https://www.naukri.com/

15 Japan Technical Communicators Association. (2019). *Technical communication symposium 2019.* https://www.jtca.org/en/symposium/index.html

16 Tekom Europe. (2019, November 7). *Technical writer: Your career in language, IT and media* [Video]. YouTube. https://www.youtube.com/watch?v=PyFY3cEd2V0

17 Google. (2018, June 29). *Meet technical writers at Google* [Video]. YouTube. https://www.youtube.com/watch?v=qnnkAWP55Ww

18 Brumberger, E., & Lauer, C. (2020). A day in the life: Personas of professional communicators at work. *Journal of Technical Writing and Communication, 50*(3), 308–335. https://doi.org/10.1177/0047281619868723

19 Davis, P. (2019, July 23). *Musical chairs: Where you sit matters.* TechWhirl. https://techwhirl.com/musical-chairs-where-you-sit-matters/

20 Lanier, C. (2018). Toward understanding important workplace issues for technical communicators. *Technical Communication, 65*(1), 66–84.

21 Jablonski, J. (2005). Seeing technical communication from a career perspective: The implications of career theory for technical communication theory, practice, and curriculum design. *Journal of Business and Technical Communication, 19*(1), 5–41. https://doi.org/10.1177/1050651904269391

22 Spataro, J. (2020, April 30). *2 years of digital transformation in 2 months.* Microsoft. https://www.microsoft.com/en-us/microsoft-365/blog/2020/04/30/2-years-digital-transformation-2-months/

23 Beauregard, T. A., Basile, K. A., & Canónico, E. (2019). Telework: Outcomes and facilitators for employees. In: R. N. Landers (Ed.), *The Cambridge handbook of technology and employee behavior* (pp. 511–543). Cambridge University Press.

24 Loh, T. H., & Fishbane, L. (2020, March 17). *COVID-19 makes the benefits of telework obvious.* Brookings. https://www.brookings.edu/blog/the-avenue/2020/03/17/covid-19-makes-the-benefits-of-telework-obvious/

25 Thompson, D. (2020, March 13). *The coronavirus is creating a huge, stressful experiment in working from home.* The Atlantic. https://www.theatlantic.com/ideas/archive/2020/03/coronavirus-creating-huge-stressful-experiment-working-home/607945/

26 European Commission. (2018). *Future of work: Future of society*. https://ec.europa.eu/info/sites/info/files/research_and_innovation/ege/ege_future-of-work_opinion_122018.pdf

27 Johnson, T. (2020, March 23). Life on reset: New dynamics emerging. *I'd Rather Be Writing*. https://idratherbewriting.com/blog/life-on-reset/

28 Flammia, M., Cleary, Y., & Slattery, D. M. (2010). Leadership roles, socioemotional communication strategies, and technology use of Irish and US students in virtual teams. *IEEE Transactions on Professional Communication, 53*(2), 89–101. https://doi.org/10.1109/TPC.2010.2046088

29 Ford, R. C., Piccolo, R. F., & Ford, L. R. (2017). Strategies for building effective virtual teams: Trust is key. *Business Horizons, 60*(1), 25–34. https://doi.org/10.1016/j.bushor.2016.08.009

30 Paretti, M. C., McNair, L. D., & Holloway-Attaway, L. (2007). Teaching technical communication in an era of distributed work: A case study of collaboration between U.S. and Swedish students. *Technical Communication Quarterly, 16*(3), 327–352. https://doi.org/10.1080/10572250701291087

31 Wiesmann, B., Snoei, J. R., Hilletofth, P., & Eriksson, D. (2017). Drivers and barriers to reshoring: A literature review on offshoring in reverse. *European Business Review, 29*(1), 15–42. https://doi.org/10.1108/EBR-03-2016-0050

32 Wilson, G. (2001). Technical communication and late capitalism: Considering a postmodern technical communication pedagogy. *Journal of Business and Technical Communication, 15*(1), 72–99. https://doi.org/10.1177/105065190101500104

33 Carliner, S., & Chen, Y. (2018, December). Job and career satisfaction among technical communicators. *Intercom, 65*(8), 23–27.

34 Bove, T. (2018, May/June). When uncertainty rears its ugly head, stretch. *Intercom, 65*(3), 40.

35 Susskind, R., & Susskind, D. (2017). *The future of the professions: How technology will transform the work of human experts*. Oxford University Press.

36 Henry, J. (2000). *Writing workplace cultures: An archaeology of professional writing*. Southern Illinois University Press.

37 Bureau of Labor Statistics (2020, April 10). *Occupational outlook handbook: Technical writers*. Retrieved November 10, 2020, from https://www.bls.gov/ooh/media-and-communication/technical-writers.htm

38 Shalamova, N., Rice-Bailey, T., & Wikoff, K. (2018). Evolving skillsets and job pathways of technical communicators. *Communication Design Quarterly, 6*(3), 14–24. https://doi.org/10.1145/3309578.3309580

39 TCBOK. (2019). *Career paths*. https://www.tcbok.org/careers/career-paths/

40 USAJOBS. (2020). *The federal government's official employment site*. https://www.usajobs.gov/

41 EURES. (2020). *Find a job in Europe*. European Commission. https://ec.europa.eu/eures/eures-searchengine/page/main?lang=en#/search

6   Technical Communication Futures

Introduction

In any profession, it is important to ensure that your skills and knowledge are current, to keep abreast of trends, and to keep up with economic and practice forecasts. These activities enable you to plot your career path, to plan your professional development activities, and to prepare strategically for your future.

A caveat before we begin: inevitably, the future is not predictable! Many of the patterns cited in the research and the practice narratives are already standard practice in some organizations, some of the predictions may already seem dated, whereas others will seem speculative in a few years' time. Nevertheless, because we explore a wide range of information sources, the chapter offers important insights into future technical communication scenarios.

After reading the chapter, you will be able to prepare yourself for future professional opportunities and challenges.

This chapter starts with a discussion of **the future of work**, examining trends in how work is conducted now and how work is changing, globally, and across professions. Many of the patterns discussed in that research are already evident or will become important for future technical communication practice. The chapter explores eight future-of-work patterns and maps those patterns to technical communication. This section includes input from the practice narratives and recent research.

Although examining the future is both interesting and strategically important, you also need to be able to perceive, amidst all the changes forecast, the **aspects of your profession and practice that are durable and those that are new**. The second section of the chapter explores competencies, synthesizing those that are constant in technical communication and also investigating new areas for competence development.

- If you are a **technical communication student**, this chapter helps you to identify work patterns and durable and new skills and competencies. It also enables you to prepare for job interviews and to plan your professional development.
- If you **work in industry as a technical communicator**, this chapter enables you to identify the extent to which your job is future-oriented and to predict and plan for the future direction of your work.
- If you are **considering a career change into technical communication**, this chapter helps you to prepare for the transition and for future opportunities.

The chapter also enables teachers to identify areas for curriculum development and researchers to see the range of research possibilities that these trends imply.

The Future of Work and Professions

Research about the future of work explores the interactions of technology with economic and social activities, particularly the trend of globalization. In a special issue of *Nature* about the future of work, Allen (p. 324)[1] explained that:

> We cannot forecast the future without an understanding of the relationships between science, technology and the economy, because technical change is such an important determinant of the future.

Susskind and Susskind[2] examined patterns of change across both new and emerging professions in their influential book, *The Future of the Professions*. They identified eight broad patterns that are relevant for all workers:

1) The end of an era.
2) Transformation by technology.
3) Emerging skills and competencies.
4) Work reconfigured.
5) New labor models.
6) More options for recipients.
7) Preoccupations of professional firms.
8) Demystification.

> Susskind and Susskind recommend that anyone trying to understand their own profession "identify those trends that already apply and to anticipate that most if not all of the remainder will take hold, sooner or later" (p. 103).

Each of the patterns they identify has an impact, to a greater or lesser degree, on the profession and practice of technical communication, both now and for the future. Some of these impacts are already evident in many technical communication workplaces. In response to each of the eight patterns of the future of work, I examine their present and future impact for the profession and practice of technical communication, based on analysis of recent research as well as online sources and the practice narratives.

The End of an Era

The concept of **Industry 4.0**, popularized in Germany, concerns the emergence of a new industrial era. It refers to a fourth industrial revolution, the era of information technology (IT)-driven manufacturing, the Internet of Things, and smart machines.[3] According to this categorization, the first three industrial revolutions were characterized by:

1) **Mechanization**. In the nineteenth century, machines replaced humans in many manufacturing processes, such as clothing and machinery production.
2) **Electrification**. Throughout the twentieth century, the adoption of electricity led to changes in how people work. Again, machines took the place of people for some types of (mainly heavy and routine) work, freeing up more time for leisure. For example, the washing machine took the labor out of doing the laundry and gave individuals, especially women, more time for other pursuits.
3) **Digitalization**. Adoption of personal computers in the 1980s led to increasingly digital, rather than physical, formats. Digital formats created new opportunities and challenges for technical communicators, including transforming how information was developed, shared, and stored. The internet accelerated the trend toward digitalization and complicated the challenges.

In this fourth era, **Industry 4.0**, work will also change as a result of artificial intelligence, machine learning, and automation.

The Implications of "The End of an Era" for Technical Communication

Because they resulted in new technologies that needed to be explained, all of the previous industrial revolutions increased work for technical communicators. Considering how these revolutions transformed technical communication practice, we can expect that this era, too, will change how, where, and with whom we work.

For technical communicators, smart machines raise several new content strategies and challenges, including how privacy and data security are handled and how content and instruction are embedded in technology. Technical communicators will document the smart machines of the Industry 4.0 era, although the documentation will not be in the old formats (manuals or online help, for example) but likely in modular, micro, and embedded content.

The Information 4.0 Consortium was founded to explore questions of how information can retain its focus on enabling people while providing content for smart machines and the Industry 4.0 era.[4]

> The Information 4.0 Consortium envisages technical communicators as user advocates in the Industry 4.0 era. The main topic areas on the consortium website are education and research, bridging silos, and the future of information.

Silos are internal disciplinary divisions, units, or departments within an organization. They are a concern of many blogs and articles that discuss the future of technical communication. The idea of bridging silos is important because product and content development are more successful when we share information across departments and sections. One practice narrative commented on how technical communicators need to learn to work more collaboratively:

> I think that most technical writers still see their role as a solo-writer job, which is not up-to-date anymore with what big companies need.

Transformation by Technology

Susskind and Susskind[2] described two kinds of transformation by technology: **automation** and **innovation**.

- **Automation** refers to situations where tasks are mechanized to the extent that they can be partially or fully undertaken without human intervention. Automation enables us to do tasks more efficiently but does not change the tasks themselves.
- **Innovation** refers to the use of technology to reinvent or completely transform tasks. Susskind and Susskind use the example of automatic teller machines (ATMs). Before ATMs, customers could get money from a bank only during daily opening hours. Although they are called automatic teller machines, ATMs did not just automate the work of bank tellers, they transformed it by offering a service that previously did not exist (24-hour access to cash). Innovative technologies are sometimes called disruptive because of their capacity to transform practice.

> An example of an innovative technology in technical communication is the wiki. Multiple authors can edit a document simultaneously without needing to manage versions, a previous impediment to collaborative writing. **Can you think of other examples?**

Since the Industrial Revolution in the nineteenth century, machines have gradually eroded the need for some jobs while creating a demand for others. Anthes[5] predicts that in the current wave of automation, some jobs will disappear, others will be created, and still others will be transformed by digital technology advances. Anthes explains that automation is now driven by machine learning, an application of artificial intelligence. Previously, machines had to be designed or programmed to complete specific tasks, but machine learning extends their capabilities because machines can "learn" to do tasks for which they have not specifically been programmed. Harari (p. 324)[6] explained how **machine learning** works:

> Computer scientists are developing artificial intelligence (AI) algorithms that can learn, analyse massive amounts of data and recognize patterns with superhuman efficiency.

Although the mechanization, electrification, and digitalization eras resulted in routine and repetitive tasks becoming obsolete, machine learning implies that even complicated cognitive tasks now have the potential to be automated.

Innovative technology transformations in the smart machine era exploit AI, robotics, data analytics, and mobile communications. **Machine learning is enabled by the extraordinary amount of data** that is available. Automation powered by machine learning will happen more quickly in fields where huge sets of data are already available.[2]

The Implications of "Transformation by Technology" for Technical Communication

Automation has already had some impact on technical communication. Writing is an activity that exemplifies how routine tasks have been automated, first through mechanization, then electrification, and then digitalization.

- The printing press automated manual text production through mechanization.
- Electrification led to some routinization of text production. It was easier to copy texts and to fix errors, for example.
- Digitalization had a huge impact on writing practice. Individual users can perform complicated word-processing tasks and save and share digital versions of their work.

All these changes led to professions rising and falling. For example, many clerical, typing, and printing jobs are now obsolete. At the same time, a huge range of digital content development roles have emerged. Through all these changes, writers remain in charge of the content we write; that task has not been automated. We do the cognitive and intellectual work of reading, processing, and analyzing other texts and formulating and producing our ideas in the form of words. **Machine learning has the potential to automate that work,** but to date it has had limited impact in technical communication. Chatbots are one example of an automated communication tool. They are often used in customer service and self-service portals. Many chatbots are rules-based and have a limited set of functions. They do not "learn" new usage. More advanced applications can perform some tasks without the need for human intervention, and they have the potential to learn through "experience" how to interact better with users.[7]

> You may have interacted with a chatbot if you contacted a customer service department through an online chat service. The chatbot performs a routine task (e.g., to manage and direct your query). It might be programmed with a set of standard questions, to which your responses determine where the query goes and who responds.

If your content is structured and you use metadata to describe the content, it can be used for chatbot content. A practice narrative explained how automation and structured content interact:

> Simple chatbots for user self-help are also on the horizon ... but all this is relatively easy now I've migrated the bulk of our content to DITA [Darwin Information Typing Architecture].

When chatbots work well, routine jobs can be automated. In a blog post for Intuillion (an Israel-based company that develops structured content), Alex Masycheff[8] explains that when routine jobs are automated, technical communicators are free to do more interesting and demanding work, including selecting output formats:

> When the boring and tedious part is done by the technology, technical writers can focus on what they do best: communicate technical concepts to the users in a way that solves [the] user's problems. This includes not just the automation of a part of the content creation process, but also allows technical writers to be more creative in choosing the channels for delivering content to users.

One practice narrative respondent saw opportunities to automate their own work:

> I'm personally preparing to better manage some specific aspects [of my work] like using software resources to automate tasks.

In the future, tasks of greater complexity could be automated. Some trends in journalism point toward possible future directions for the automation of repetitive writing tasks in technical communication. For example, where the reporting is routine, capable of being produced by a machine, and the most important information is variation in data (e.g., financial indices and sports results), automatic reporting is already prevalent in journalism.[9] Even if writing can be automated, content will still need to be selected, organized, and structured, and technical communicators will continue to play a pivotal role, as Seth Earley explains (p. 14):[10]

> Amidst the appeal of new technology, technical writers serve a vital role in development of AI applications, including chatbots, by being the subject experts most equipped to address the content requirements.

Automation has ethical implications that also need to be considered. Although it should foster inclusivity, automated content, without adequate attention, can perpetuate existing social biases and stereotypes.[11] This is a compelling subject for future technical communication research.

Emerging Skills and Competencies

Susskind and Susskind[2] outlined several new and emerging skills that have an impact on all professions, including technical communication. These include the following:

- **Different ways of communicating**: Old communication technologies, like the fax machine, are now largely obsolete, and new technologies, like social media and chatbots, have emerged. As remote working becomes more common, face-to-face meetings are gradually being replaced by virtual meetings.
- **Mastery of data**. Every application you use gathers usage data of some kind. Being able to use and analyze those data is emerging as a required professional skill. Because most online tools and applications now gather usage and user data automatically, datasets are increasing in a phenomenon known as **big data**. Being able to analyze huge datasets gives us insight into users and patterns of use. Analysis of big data can help us to develop personalized content.
- **Diversification**: As we have seen, diversification is a considerable trend in technical communication; it is a trend in all professions[2] and has emerged because interdisciplinary practice is necessary to meet clients' requirements. Your content users do not care about your job title or professional discipline, however. They just want help using information, a device, service, or application.

The Implications of "Emerging Skills and Competencies" for Technical Communication

The World Economic Forum's *Future of Jobs Report 2020* had positive news for content industries, including technical communication. It signposted content production among a "set of emerging professions [reflecting] the continuing importance of human interaction in the new economy" (p. 8).[12] In order to be effective, however, we, like other professionals, need to keep learning and upskilling. Technical communicators have always had to learn new tools and **new ways of communicating** and this trend will continue. Mehlenbacher[13] described technical communication as a means "for mediating between technology and humans" (p. 195). As long as that technology continues to evolve, our work does too.

In many industries and sectors, data are used to target customers and to promote products. Hoffman (p. 8)[14] explains how **mastery of data** can improve technical communication:

> Metrics like page views, time on page, click-through-paths, knowledge journeys, search terms, meta tags, and ratings are insights that inform you on how your customer is not only consuming your information, but how they are experiencing it.

Technical communicators will need to be able to gather, access, understand, process, analyze, and use data to target and customize user content effectively.

Ethical concerns have been raised about how big data are used. In many parts of the world, data regulations describe how data should be gathered, stored, processed, used, archived, and deleted. For example, in the European Union, the General Data Protection Regulation came into force in May 2018.[15] It has quite strict regulations about how data are gathered, processed, and managed. Data breaches incur severe penalties. Technical communicators need to be aware of the data protection regulations in their markets.

> Because technical communicators are user advocates, **understanding and promoting ethical data collection and analysis** may be an increasingly important competence for us in the future.

In technical communication, **diversification** has been a big issue for over two decades. It appears to be driven, at least in part, by a client-focused and personalized approach to content delivery. For example, Hoffman argues that diversification away from the "old school" technical writer role is needed and that a new role, Director of Content Experience, is emerging because of the impact of technology changes and the need to "focus on the customer first and foremost" (p. 8).[14] Some practice narratives indicate that the trend of diversification is likely to continue, as these excerpts explain:

> Lines [are] also blurring between jobs so you'll now find technical authors in a variety of jobs (definitely writing) that they wouldn't be involved in in the past (e.g., QA [quality assurance], business analysis, etc.).
>
> For technical communication in general, I worry that in the rush to "specialize", we've diluted what we do. Each little niche has its own job

title (content strategist, information architect, e-learning developer, user assistance developer, technical author...) which is good in one way, but they all overlap considerably. I'm a generalist, and it's what I love about what I do: the variety.

Although diversification has been a trend for a long time, it may lead to fragmentation and a loss of professional identity. Ensuring the recognizability of this profession is an important strategic goal for the future.

Professional Work Reconfigured

Susskind and Susskind[2] described three ways in which work across all professions is reconfigured because of technology:

Routinization: Some tasks can be reduced to a repeatable procedure by technology (e.g., through templates, standard online forms or checklists). These routines help us to be more efficient and to spend time on work that is more interesting. This type of reconfiguration interacts with automation.

Disintermediation and **reintermediation**: Brokers help clients to access many services. For example, they may help you to find the cheapest service available. Many such intermediaries are no longer needed given that nowadays individuals can access services directly online. This trend is known as disintermediation. It has become widespread as the internet has replaced many brokerage services.

> For example, when you book a holiday, you may conduct your research online and make direct bookings with an airline and hotel instead of booking the package through a travel agent, as was common in the pre-internet era.

In some cases, accessing services directly is slow, inefficient, and ineffective. Reintermediation happens when experts take back the role of interfacing between a client and a service provider.

Decomposing and multi-sourcing: these terms refer to the common corporate practice of dividing work into tasks and projects, often distributed across teams in various locations and outsourced in some cases. For most technical communicators, this type of labor division is already common.

The Implications of "Professional Work Reconfigured" for Technical Communication

In technical communication, some editing tasks are **routinized** (e.g., by grammar checkers). Four practice narratives mentioned Acrolinx,[16] an automated editing tool. One practice narrative acknowledged that automation of routine tasks could help them to take on a more intellectually challenging role:

> There will be more automation and my role will become less about producing documents and more about having a deep understanding of the English language and editing and reviewing the language used by developers.

Disintermediation has resulted in a situation where users of technology and information systems have more direct access to organizations (e.g., through social media) and can propose changes[17] or even contribute to content development (e.g., in user-generated content). Another type of disintermediation is when user assistance is embedded in a product. Instead of relying on a manual or help file, the user gets assistance directly from the product.

Reintermediation happens when technical communicators take back ownership of content. For example, if user-generated content is incorrect or misleading, a technical communicator who curates that content acts as an intermediary by updating the content or fixing its errors.

New Labor Models

Technology has led to many new ways of working. Susskind and Susskind[2] listed several new models, including the following ones that are relevant for technical communication:

Labor arbitrage. This term refers primarily to the practices of offshoring and outsourcing, both discussed in Chapter 5.

Flexible self-employment. The so-called "gig economy" describes how workers connect with a service provider (nominally, their employer) to get short-term, immediate work, or "gigs." Examples of work in the gig economy include food delivery, accommodation letting, and taxi services. Two features of this labor model are ratings-based feedback and online payments through the service provider. For example, when you book an apartment through an apartment rental platform, you pay the service provider, not the individual who owns or lets the apartment. A further feature, which may benefit both the consumer and the employer by keeping prices low, is an oversupply of workers.

To work in the gig economy, you need basic digital skills, an internet-enabled device, and the means to offer the service (e.g., a bicycle for food deliveries in a city, a car for taxi services, and an apartment or spare room for accommodation letting). Workers have flexibility in deciding what "gigs" to take, but the cost of this flexibility is that the gig economy by its nature is precarious. Furthermore, workers have limited or no access to the benefits of health insurance, holiday pay, and sick pay that directly hired employees expect. Workers depend on good feedback, and they are at the mercy both of the individuals who use their service and of the platform through which the service is offered.

Users: The open-source movement was an early model for how individual users became contributors to professional services, instead of just clients. Susskind and Susskind explained that "non-professional lay people are emerging as a new source of practical expertise" (p. 128).[2] In journalism, for example, this phenomenon is known as citizen journalism, where members of the public report or contribute to news stories, a development facilitated by technology, particularly by smartphones and social media.

The Implications of "New Labor Models" for Technical Communication

In technical communication (as outlined in Chapter 5), jobs continue to be advertised primarily on a permanent, direct-hire basis. Three trends are emerging and might become more dominant, especially during periods of economic uncertainty:

- Chapter 5 outlined a trend toward **contracting companies** offering technical communication services. These companies seem to be more successful during economic downturns because they pick up work contracts in companies where technical communicators have been laid off.
- The second trend is a more explicit example of the gig economy, where technical communicators advertise **freelance services** for short-term contracts through applications like Upwork.[18] On Upwork, each freelancer has a profile that outlines their services, skills, preferred project time frames, and hourly rates. Each profile also includes a "job success" percentage, based on feedback from previous work. The service user inputs project criteria and is matched with available freelance technical communicators, who choose which gigs to undertake. This type of service enables freelancers either to undertake enough work to fully support themselves or to supplement another income.
- The trend of **user-generated content** has been prevalent in technical communication for more than a decade. This content is generated when a user contributes to a public forum (e.g., a product discussion forum) to explain how to use a feature or carry out a procedure. Some technical communicators already curate, or expect to need to curate, this type of content. One practice narrative noted that:

  > Managing customer input is increasingly likely to feature in the responsibilities of a technical writer.

More Options for Recipients

Clients, the recipients of professional services, now have more ways to access services (e.g., by selecting services online, collaborating online, accessing personalized content, using online self-help platforms, and accessing embedded content). Susskind and Susskind[2] explained how these service delivery options have affected all professions. In technical communication, their impact is significant.

The Implications of "More Options for Recipients" for Technical Communication

Technical communicators respond to the drive to offer clients more access to services and information in several ways.

The move toward **self-service** information platforms is already prominent in technical communication. Technical communicators are responsible for developing and publishing the content that users access online.

Personalization is becoming an essential characteristic of this type of content. In an *Intercom* article, Scott Abel explains personalization as "getting the right pieces of content in front of the right person, at the right time, in the right place, in the format that individual requires, and in essential ways that provide value to them" (p. 5).[19] Personalized information delivery means making sure the user gets information targeted specifically to their needs, sometimes when they do not know their needs. Users now expect personalized content, and platforms like Amazon and Netflix are examples of content providers that do personalization well.[14]

> If you have made purchases on a site based on its recommendations, that is an example of effective personalization.

Content that is tagged clearly, with metadata that describe its purpose, potential uses, and audiences, facilitates personalization. Through structured content (e.g., using DITA) and metadata, content can be dynamically delivered to users on the basis of individual searches.

Preoccupations of Professional Firms

In the research for their book, a set of "preoccupations" dominated Susskind and Susskind's[2] discussions with professionals. These included the following:

Liberalization: the opening up of the professions to new fields of work as well as reducing the monopolies that professionals exercise (a concern I examine with respect to technical communication in Chapter 7).

Globalization: No industry or profession has been unaffected by globalization, the trend toward "flows of people, goods and information around the world" (p. 136).[2]

Specialization: Just as diversification is a trend, so too is specialization. It is often necessary, in order to distinguish yourself in a labor market, to become an expert in a niche aspect of practice.

New business models: As more of us work from home and work is less closely monitored by managers, it makes sense that value is placed on "output rather than input, for the value delivered rather than the effort expended" (p. 137).[2]

Restructuring: In the recent past, most professionals, especially those in the traditional professions like law and medicine, worked as sole traders or in small organizations. It is difficult for small firms to compete with larger entities, in either pricing or service offering. This situation has led to a culture of corporate acquisitions and restructuring.

The Implications of "Preoccupations of Professional Firms" for Technical Communication

Globalization impacts technical communicators in many ways. We work with individuals from around the world. Technical communication roles are advertised in many regions of the globe, and our content is translated into multiple languages and localized for content users globally. Translation has not always integrated seamlessly into content development life cycles, but some practice narratives suggested ways in which this integration might improve in the future:

> Translation is the biggest pain point especially with continuous delivery of products. Hopefully with improvements in machine translation, this will become less of an issue.
>
> [I am working on] creating professional tools like a company termbase to optimize translation costs and time.
>
> [More] video [and] graphics will be used (less need of translation).
>
> [I hope we can] improve how we manage translation processes with translation providers.

Most technical communicators **specialize** to some degree, differentiating ourselves through the tools we use or the domains we work in, for example. Many practice narratives outlined ways in which the respondents specialize:

> Even within software teams in a company, people are quite specialised.
>
> I'm one of a handful of authors that specialise in high tech [hardware] that comes with [software].

It is important to choose your specialism, rather than fall into it, and to be strategic about how that specialism matches your career goals. Specialization can be negative when it limits your career opportunities, as these extracts show:

> Sometimes I think I have pigeon-holed myself... In my experience, this niche role has a salary limitation and I know I have hit the ceiling.
>
> I'd advise anyone to really plan their career. I did the things I was good at and was asked to do until they became my specialism, but I realized, too late

I think, that my peers were strategically moving into higher-level positions with less work while I was doing the work they didn't want to do.

Chapter 5 examined the impact of **restructuring and acquisitions** on technical communication workplaces. This corporate practice can bring opportunities and challenges. It may lead to greater visibility and more interdisciplinary and multidisciplinary teams. At the same time, it can be confusing and destabilizing for individual employees. A couple of practice narratives suggested unease about restructuring and its impact on future work opportunities in the profession:

> The next few years are hard to predict, aside from possibly finding work as a content writer (in the non-technical domain) or something similar.
>
> Unfortunately, the outlook for my job is not positive. The older product has had development stopped for 12–14 months until bugs are fixed, which means there is little work for the technical writers to do. My newer product will be made obsolete by an acquisition in the next few months, which also means we literally don't know if we are needed going forward.

Demystification

Until the late twentieth century, traditional professions (like law and medicine) were regarded as somewhat mysterious and unreachable, "outside and above the working class" (p. xvi).[20] Professionals sometimes encouraged that sense of mystery by using jargon and by complicating straightforward processes. As contemporary work is broken down into tasks and as services and information are available online, the mysterious aura of many professions has begun to evaporate. Furthermore, we can conduct research and learn online, often for free. This access to information makes us powerful and it simultaneously decreases the power of many professions.

The Implications of "Demystification" for Technical Communication

Technical communicators, by producing clear information, are part of the movement to demystify products and services. Technical communicators who produce e-government content, for example, give citizens access to public information and services. As the era of traditional professions may be on the wane and professions have been demystified to some degree, new and emerging professions have scope to emerge and develop. Now may be a good time for technical communication to position itself, along with other new fields of work. As an information profession, we have many opportunities.

The Implications of Covid-19 for the Future of Work

Although Susskind and Susskind's patterns help us to catalogue the ways that work has changed and is changing as a result of technology and globalization, the events that have the most impact on our work, social, and personal lives are often the least predictable. At the time of writing (late 2020), the world remains in the grip of Covid-19, the disease caused by the novel coronavirus. With its associated lockdowns and restrictions on movement and travel, this set of circumstances has overwhelmed news cycles to the exclusion of most other events. There is just one news story, globally.

Covid-19 has accelerated the future of work, for the short term at least, changing how we work, learn, and socialize as no previous event in our lifetimes has. Furthermore, many of the changes it has brought about will likely last beyond the life of the virus and may inspire new movements. For these reasons, it is worth considering how this global event may affect work in general and technical communication in particular. In the midst of the pandemic, it is difficult to process the current impact or predict future developments in a meaningful way. Nevertheless, some sources[21,22,23,24] have suggested how the pandemic will impact the future of work. The main predictions these sources propose for organizations are that:

- Some **businesses will fail** as a result of the prolonged closures that the pandemic has necessitated, thus changing the labor market. Certain industries, including hospitality and travel, are acutely affected.
- Other **businesses and sectors will thrive**; these include those offering online communication solutions (e.g., Zoom was an early success story).
- Accelerated digitalization may cause **structural changes in economies**, leading to "temporary disruption in the labour market as well as skill mismatch" (p. 25).[24] Not enough people will be qualified for new roles that emerge; at the same time, many individuals in retail, hospitality, and travel industries may lose their jobs and need to retrain.
- **The value of essential workers** will become evident, and frontline workers, who were often among the lower paid and most precarious employees, may be better compensated.
- As more people work successfully from home, the need for **corporate real estate** may be re-examined.
- Organizations will have to **plan for resilience** in case of future pandemics.

Individual workers may find themselves:

- **Working remotely** some or all of the time or working staggered shifts to ensure that workplaces can facilitate social distancing.
- In a more **competitive job market** because of business failures.
- With **less need to travel** for work, as more meetings and events that pivoted online continue to operate online.
- **Seeking work with a higher purpose**, based on re-evaluating their work's purpose during the pandemic.
- Needing to be more **adaptable** to thrive in a less predictable work environment.

Some of these changes have already begun to happen, others are aspirational, and many will be short-term.

The Implications of Covid-19 for the Future of Technical Communication

When I gathered the practice narrative data in late 2019, Covid-19 was as yet unknown and therefore does not feature in the narratives. Furthermore, we do not know the impact of the pandemic for the medium or long term. Bloggers, podcasters, and other technical communication influencers have begun to analyze the situation and strategize for the future on the basis of their predictions for the impact of Covid-19 on technical communication. These sources include the following:

- Survey data, from a survey run by Tom Johnson in May 2020 about the impact of quarantine and the pandemic on technical communicators.[25] He received over 270 responses from individuals in almost 40 countries, and about 50% of respondents were based in the US.
- A podcast by Ellis Pratt from Cherryleaf[26] exploring the topic, with reference to several news sources, from April 2020.
- Kirk St.Amant's Covid-19 essays.[27]
- A Write the Docs Australia webinar that featured a conversation about technical communication and Covid-19 between Tom Johnson and Kirk St.Amant, followed by a discussion.[28]

According to these sources, the key impacts will be on:

Employment: Over 90% of Johnson's respondents were still employed, and more than 80% had not experienced a salary reduction. Nevertheless, responses suggested that recruitment of technical communicators would slow down for the short-term; 53 respondents indicated that recruitment of new hires was cancelled or on hold, and eight provided qualitative feedback that pointed to a similar trend.

Remote working: As discussed in Chapter 5, remote working was already an option for many technical communicators, and this trend is now a generalized work norm. Almost 50% of Johnson's survey respondents predicted that remote working would become more acceptable as a long-term consequence of the pandemic.

Opportunities for some software/applications: Zoom, an online meeting tool, moved from niche to mainstream within a few weeks. Many other tools that facilitate remote communication have had a similar boost.

Increased need for technical communication: During lockdowns, online self-service platforms replaced some types of user support, like call centers. At the start of the pandemic, call centers were overwhelmed and waiting times were untenable, as individuals needed to make or change plans at short notice (e.g., for travel, insurance, home technology, infrastructure, and health care). At the same time, it was difficult to accommodate social distancing in busy open-plan call centers, or to set up work-from-home protocols for employees, because of technical difficulties, potential security risks, and infrastructure limitations. Self-service portals made sense because if users could be directed to access information online to solve their queries, then the call queues for those who needed to speak to an agent were much shorter.[26] If users have a good experience with online portals, this shift toward self-service may become a long-term consequence of the pandemic and ultimately will increase the need for technical communicators.

Remote learning: In a related trend, most education and training offerings switched to online delivery during Covid-19. This change brought about a need for instructional design and online training expertise. Technical communicators have this expertise and may have more scope to work in training and education development as a result.

Volunteering: St. Amant[27] described various ways in which communication professionals can contribute to their local communities by developing materials that offer clear explanations of central messages (e.g., how to recognize symptoms of the virus or how to care for others who have contracted the virus). Individuals need clear and direct instructions to know how to behave in changed circumstances.

Trustworthiness: Users need to be able to trust online content. If people have to use a portal rather than interacting with a person to access information, they need to have confidence in that information. In the same vein, people need to be able to trust public health information.

Durable and New Competencies

In any profession, a set of core competencies will always be necessary. These durable competencies enable us to identify what is distinctive about our profession. Given the foregoing discussion of patterns that are emerging in some areas and the matching decline in others, it is also likely that some new competencies will be needed in the future.

Durable Competencies in Technical Communication

Throughout the development of technical communication as a profession and practice, certain competencies have been essential. This work combines a whole suite of competencies, discussed in detail in Chapter 4 and broadly summarized as **the ability to communicate, explain, and help people to learn, understand, use, and interact with specialized concepts or technologies**. These competencies incorporate writing, communication, visual design, user advocacy, and affinity for technical subject matter. Although our rhetorical work is essential, we also need to understand the possibilities and affordances of technologies. Skills with individual applications date rapidly, in as little as five years,[29] but **using technology to complete core tasks** is a durable competence.

Box 6.1   Practice narrative excerpts about perceptions of adaptability

The main thing is to keep learning, keep being curious, don't be intimidated – just chip away at a new subject, and it will start making sense.

My job will be subject to more and more changes ... for the tools and methods I'll have to learn and use. Innovation will impact it a lot and more skills and capabilities shall be implemented to meet customers' needs and expectations.

We need to adapt to the changes in media as we steer away from paper to multimedia format.

The acceleration of deliveries (continuous delivery) driven by the market today is increasingly adding complexity to my job.

Preparation is hard as where it's going is very fluid and the only real thing you can be is adaptable.

An additional competence that has been required since at least the 1980s is **adaptability** to deal with changes, particularly in relation to technology. Greg Wilson[30] described the "postmodern" work of the technical communicator, as many of us adapt to continual change. Therefore, **being flexible and responsive** may be the most durable and important competence for your professional future. Several practice narratives alluded to adaptability in changing environments. The excerpts in Box 6.1 are just a sample of the many narratives about the future of technical communication that mentioned change or adaptability.

New Competencies in Technical Communication

Several sources point toward new areas of growth in technical communication. Although these growth areas continue to build on your existing writing, technology, and rhetorical competencies, they also suggest directions for your future career. New or emerging competencies that practitioners cited in the practice narratives are illustrated in Table 6.1.

Table 6.1 New competencies: summary of practice narrative responses

Competence	Number of responses
Creating video instruction	8
Developing intelligent content	7
Writing microcontent	6
Writing application programming interface (API) and developer documentation	4

Creating Video/Multimedia Instruction

In a 2005 survey of technical communication managers,[31] only 4% of respondents expressed a need for video content developers. In the years since that survey, platforms like YouTube and TikTok that enable users to upload video content have led to a huge rise in user-generated video instruction. The popularity of video content responded to a latent demand. Users did not know they needed video instructions until user-generated video became widely available. This development also created a **need for professional video content** that gives ownership and control of the instruction back to organizations.

Video and multimedia development have been required competencies in some technical communication roles for many years. Nevertheless, these were identified as future competencies by eight practice narrative respondents, suggesting that these competencies are still new or challenging for many technical communicators. These excerpts capture the thrust of the comments:

> I already do computer-based training materials (Captivate simulations, for example) but can see this moving more and more to video. The popularity of YouTube shows this trend.
>
> Diversification of the deliverables: the means to convey information need to be more appealing [than] a long technical document (video, digital guidance, etc.).

Some narratives were cautious about how to deliver video effectively:

> Videos and more attractive walkthrough fits more to customers' needs but making video tutorials for the moment is a big big job and companies do not seem to have the necessary skilled [people].
>
> There is always a lot of talk about how users like videos, but my colleague and I have really struggled to find the time. We never really have time to spare after doing the basic documentation – which is essential for a large and complex product like the one we document.

These excerpts show that technical communicators need support and training to be able to create effective and professional video content. They also imply that academic and professional development instructors need to create stronger multimedia and video capacity in our programs.

Developing Intelligent Content

As discussed in Chapter 4, topics have replaced manuals as the dominant content format for most types of technical communication over the past two decades. The persistent move toward online, short-form, structured content will continue.

> Content for the Industry 4.0 era is dynamic, molecular, offered (rather than delivered), spontaneous, and profiled (or personalized).[4]

In a *Content Wrangler* webinar, Alexander Hoffmann[32] explained an emerging, and related, competency: **developing intelligent content** for smart machines.

Intelligent content is "structurally-rich and semantically-categorized, and is therefore automatically discoverable, reusable, reconfigurable, and adaptable" (p. 33).[33] These features mean that intelligent content can be **personalized**, depending on the features of the user's context (e.g., the device or version they are using, the task they are undertaking, specific configurations or settings, recently used features, and recent searches). As we discussed earlier in the chapter, users now expect content to be personalized. That said, personalization is often random and inaccurate. Targeted advertising, for example, can respond to a single search (and one that is not relevant to our usual interests or indicative of a pattern).

Developing intelligent content has several related or sub-competencies:

- Developing reusable content.
- Using metadata and structuring content.
- Managing the life cycle of content.
- Managing content components.

Developing reusable content. In topic-based writing environments, single sourcing has been the way to achieve reusability and has been a goal for two decades. In tandem with the related trends of topic-based, modular content, single sourcing had the potential to transform technical communication practice. Nevertheless, the implementation of single sourcing to date has been neither seamless nor transformative. As Mark Baker (author of two books on structured/topic-based writing) argued in a blog post,[34] "single sourcing has been the watchword of technical communication for the last several decades" but "we have never fully made it work."

A practice narrative explained the widespread use of, but also problems with, single sourcing:

> the writing industry tends to go through "buzz" phrases – currently single-sourcing is very much 'in' but, as in all cases, it involves compromising to save costs (neither, e.g., [PDF]/online is the 'best' it could be) and only really benefits larger companies that can pay for the best compromise design.

As this excerpt shows, single sourcing works most effectively in large organizations.

> Bloggers, including Mark Baker, Sarah O'Keefe, and Alan Porter, have recently suggested that compared with single sourcing, using a single publishing system (with multiple sources) is more realistic and may make reusability more manageable.

Although structured content is not a new concept in technical communication, several practice narratives explained how respondents are currently moving, or have only recently moved to, a structured, usually topic-based, system for content development:

> Moving from monolithic documentation to topic-based authoring.

Another narrative explained how the technical communicator of the future:

> Need[s] to have expertise in using content management systems, including basic skills in coding (at least XML and HTML). More importantly, they will need to be trained in writing content that fits content management strategies (like topic-oriented writing and single-source approach).

Using metadata and structuring content: In order for content to be reusable, it must be retrievable and that is achieved by marking up content and adding metadata.

Managing the life cycle of content. Technical communicators often work as content strategists or help content strategists from planning through to archiving content.

Managing content components. Increasingly, technical communicators develop components rather than long documents and these components need to be managed. In order to effectively manage content and content components, technical communicators need to select appropriate tools and systems. At the moment, we have a plethora of tools for creating and managing structured content. Large organizations have invested time and money in adopting DITA. Therefore, it seems likely that they will continue to use it in the near future, and it will continue to be updated. The current DITA specification, 1.3, was published in 2018.[35] Although DITA offers large organizations consistency and efficiency, its disadvantages include complexity, of both the architecture and its implementation. DITA is most beneficial for large technical communication teams, but even in large teams, it takes a long time to learn and to implement. It is also difficult to teach DITA in an education setting, and many students enter the workforce with only a theoretical understanding of structured content and single sourcing. In Lanier's survey[36] about work issues in technical communication, DITA and related aspects of content management were mentioned by 39 respondents (23%), indicating that the transition to structured content is a live issue for practitioners.

Some alternatives to DITA were discussed in Chapter 4 and mentioned in practice narratives and other sources. These include lightweight markup languages like **Markdown** and **AsciiDoc**. Alternatives that may be adopted more widely for intelligent content development include the following:

- **Lightweight DITA:** a version of DITA that "has a smaller element type and attribute set, stricter content models, and a reduced feature set."[37] A key feature is that it has three authoring formats: XML, Markdown, and HTML5. It was published by OASIS in October 2018. It is too early to say to what extent it has been adopted in industry, but it may become an alternative to DITA in situations where DITA would be too complex to implement.

- **Schema.org**: an alternative means of structuring content, not mentioned in the practice narratives but supported by several corporations, including Google. Schema.org is "a collaborative, community activity with a mission to create, maintain, and promote schemas for structured data on the Internet, on web pages, in email messages, and beyond."[38] Schema.org enables you to structure website content to increase its findability and works on the principle of marking up content (using HTML5 microdata tags) to add semantic information to HTML code. The microdata enable you to provide semantic information to search engines about the topic. Schema.org is similar to DITA in distinguishing content "types," one of which is "how to." The **How to** schema is likely to be especially useful for technical communication because it offers a means of presenting procedural information that search engines can recognize, leading to improved search results.

These alternatives may already be in use by technical communicators who write developer documentation. Regardless of the tools we use, it is likely that concepts such as content reuse and content structure will prevail. Carlos Evia made this point in a 2018 interview with Tom Johnson:[39]

> DITA might not be here in 2030, but I want to think that intelligent content and reuse will still be important for the new gadgets and devices that increasingly diverse and demanding audiences will use in the future.

Writing Microcontent

The movement toward embedded content is leading to a parallel rise in the need to be able to **develop microcontent**. Microcontent has two possible meanings:

1. Nielsen and Loranger[40] use this term to refer to headings and titles that help readers to understand the focus of online content and to scan.
2. In technical communication, microcontent refers to the move toward creating short pieces of content that are often embedded in a device or application.

Several practice narratives discussed the increasing relevance of embedded content, as in these excerpts:

> Eventually I think there will be a switch from trying to deliver tailored documentation in different versions at an accelerated pace, to having more and more embedded content within the product itself, simplifying the user experience.
>
> As far as the job itself is concerned, I believe that the need for truly responsive documentation taking advantage of the features of smart devices will grow.

Topics replaced manuals and evolved from information mapping, minimalism, and online help. In the future, topics may be replaced by microcontent as the new dominant format. Hoffman explained this evolution from manuals to microcontent (p. 7):[14]

> Printed user manuals moved onto a desktop screen, the desktop turned into a laptop, the laptop into a smartphone, then into a smartwatch, washing machine, VR [virtual reality] headset, or even an Alexa on your customer's night stand. We need to stop "writing manuals" and start crafting usable, findable micro-content.

The Information 4.0 Consortium describes the content of the future as "molecular" rather than modular, signifying this shift from topics to even shorter content chunks.

Microcontent can be adapted and reconfigured to be embedded in interfaces and smart machines or used by chatbots.

Some excerpts from the practice narratives suggest this shift in focus from topics to snippets:

> The job of a technical writer will include more microcopy jobs.
> Micro content will become much more significant as users demand concise targeted information for an immediate need.

This trend is linked to a move toward user experience (UX) writing. UX writers create the text that guides a user through a user interface (UI). The vignette in Chapter 2 saw June returning to education to complete a master's degree with a UX component so that she could move into a UX writing role. Two narratives focused on how UX writing has displaced technical communication:

> Many of the tasks that used to be the sole preserve of software tech writers are no longer needed or are now performed by other professionals, such as UI/UX designers or software developers. This trend could continue, so the number of interesting, well-paid documentation roles for somebody with my level of experience might reduce sharply.
> User experience research and design will do away with a lot of the low-level technical writing jobs.

One practice narrative argued that technical communicators need to take ownership of the shift toward embedded content:

> As more and more tasks are being mechanised and controlled by software, more and more of the instructions to accompany them are being embedded

within that software. It's another form of technical communication - and we need to own it. If we don't, another "specialism" will form – "user contextual support advocate" or some such, and the job will be done by someone who perhaps has many skills but won't be seen as a technical communicator.

One respondent focused on the positive impact of aligning with UX professionals:

My main tasks already include writing and reviewing interface content and other UX writing. There is huge potential for us in this space.

Lebson[41] listed technical communication among many UX careers, matching this respondent's belief that UX writing presents an opportunity rather than a threat for technical communication.

Writing API and Developer Documentation

The ability to write application programming interfaces (APIs) and developer documentation more generally are growth areas in technical communication. APIs enable web applications to interact with each other. Developers use them to add functionality to websites, but they need API documentation because APIs are often difficult to use. Technical communicators write that documentation. In 2020, Tom Johnson[42] surveyed technical communicators who focus on developer documentation. His findings show some trends in this sector:

- Static site generators (like Jekyll, Hugo, Gatsby, and MkDocs) are the primary authoring tools.
- The docs-as-code approach to writing (treating documentation in the same way as developers treat code) is popular in this sector.
- Markdown is the most common documentation source format.
- HTML is the most common output format.
- Most developer documentation is not localized.
- The most common form of developer documentation is Representational State Transfer (REST) APIs.
- One third of respondents were former engineers or had come from a highly technical role.
- Over one third of respondents were lone writers, and most (over 85%) were full-time employees.

Four practice narratives predicted an increased demand for API documentation and developer documentation more generally. Writing developer content or content for highly technical domains may require **the ability to**

understand and use coding languages. These excerpts explain the trend and its implications for technical communication:

> More internal documentation, rather than only documentation for end user: e.g., technical processes for development teams. I have to learn some basic code myself: e.g., SQL.
>
> I think documenting APIs may be growing. APIs are a great way for engineers to combine existing off-the-shelf software with what they're doing. In turn, they can sell/offer their software to others. Technical writers are needed for this.
>
> The only change I foresee for myself is trying to define a new way of handling API documentation: developers aren't satisfied with just Doxygen comments, and I'm investigating how they can produce something where we can complement this basic info with more context, examples and so on.
>
> Change: More and more demand for API-conversant tech writers; Prep: Learn about API; resume battle with Python.

How Technical Communicators are Preparing for the Future

I asked practice narrative respondents to discuss how they were preparing for their future at work. Although the responses varied greatly, four themes emerged: stretching competencies, professional development, strategizing, and mentoring. Two respondents indicated that they expected their role to be unchanged in five years and therefore they were not planning for the future. Three others indicated that they were coming close to retirement and therefore had no need to plan for their role into the future. One respondent planned to forge a new career in an area unrelated to technical communication.

Stretching Competencies / Trying New Areas of Work

Many of us find ways to develop and stretch our skills in our current work environment, such as volunteering or joining committees. If there is a particular aspect of work you would like to know more about, it is essential to seek out opportunities to work in that area. This enthusiasm does not go unnoticed and often is rewarded. For example, you might get to do more work in an area you have expressed interest in.

One practice narrative respondent recommended "trying to strengthen your experience by taking on more responsibilities and challenges at work", and another individual was stretching their skills by piloting a project in GitHub to manage and publish technical information as a response to the trend for technical information to be "more integrated with SW source code management (e.g., in Git)." Another had joined a committee and needed to read about company policies to be able to contribute.

More Education, Training, and Professional Development

Ten respondents mentioned their plans for more training and professional development of various kinds, much of it informal, like reading, keeping up with trends, or trying out new techniques and applications.

> I need to start reading a lot more around current industry practices.
> I will continue "upskilling" and observing trends in technical communication so that my work and my role remain relevant.
> Learn how to create interactive images, nested dropdowns within Q & A's, and help content that has a similar 'journey' to installation wizards.

In two cases, respondents mentioned that they were planning to engage in training as a strategic response to perceived employment uncertainty:

> Engaging in training and education is necessary so that I am ready if I need to move on again.
> It looks like I would need to retrain if I couldn't get recruited as a technical writer again.

The narratives imply that, as a practitioner, you have some agency to decide what areas to focus your attention on for training and development, but to quote one, and paraphrase many, "finding the time is always a challenge."

Strategizing for the Future

Respondents reported on strategizing activities they were undertaking. At a personal level, these activities included time management, networking, and career planning. Three respondents hoped to move into management roles:

> I've been thinking about how to evolve to managerial level.
> I anticipate moving back into management, so I can blend my technical writing skills with my abilities to manage projects and to recruit and manage people.
> I might have to pivot to [management] track if only to keep my income at a certain level.

At a team or organization level, strategic initiatives that practitioners were planning included analyzing teams, systems, and competitors.

> I'm trying to work out new ideas about how a technical writing team should work, for example applying concepts from Agile approaches (like proactivity, focus on customer needs, specialization vs. collaboration).
> Keeping one ear to the ground to be ready for the changes that will come (after 35+ yrs. in the business I know that).

Mentoring Colleagues

Three respondents reported activities they were undertaking to mentor colleagues and to have a career legacy. The examples below show that mentoring may take the forms of offering career advice, helping new hires to find their feet, and acting as a sounding board for colleagues:

> When I get a new hire, I have to remind him/her to cultivate their career and not stay too long in [a] small company.
>
> My plan is to keep listening to clever people's ideas, share mine, get feedback, and refine.

Of course, some practitioners were undertaking multiple activities to be prepared for the future, as these excerpts capture:

> I'm preparing for all the above by studying, implementing competencies, updating my knowledge, sharing know-how, increase[ing] relationships and new contacts with experts.
>
> Preparation is more classes, more reading, more professional networking.

There was no indication in the practice narratives that participants were strategizing at the level of the profession or exploring ways that they might contribute to the profession in the future.

Summary of Chapter 6

The future of work and the future of the professions are active research areas. This chapter examined eight patterns of change in traditional and emerging professions.

- Many future-of-work patterns already influence, or will influence, technical communication. They include emerging skills, new labor models, and technology transformations.
- Covid-19 has had sweeping and potentially long-term impacts on work in general and technical communication in particular.
- The future of the technical communication profession will be shaped by Industry 4.0 and Information 4.0, technology developments, and how we respond to those changes.
- The future of practice will be influenced by single sourcing and structured content models, the movement toward personalization and microcontent, and greater need for developer documentation.
- The chapter reveals several durable competencies, particularly adaptability and writing and rhetorical skills. New competencies include various skills associated with writing intelligent content, coding, and developing video-based instruction.

- The practice narratives reveal that technical communicators are aware of many changes in their profession and practice, and they are planning for their futures by engaging in professional development, stretching their existing skills, and undertaking strategic activities such as networking and career management.

Discussion Questions

1. Based on your reading of this chapter, what predictions for future work trends do you think are most exciting? Which ones have you already experienced?
2. Identify examples of automation, and of innovation, in technology. Can you think of examples of automation and innovation in the technical communication profession?
3. The future of technical communication is discussed regularly in blogs and magazines. Find a recent article about the future of the profession. What trends does the article predict? How are these predictions similar to and different from the trends discussed in this chapter?
4. To learn more about how Industry 4.0 and Information 4.0 will change your work, explore the Information 4.0 Consortium's blog.
5. Which work patterns discussed in the chapter were you already aware of? Which ones do you want to learn more about? Find sources that can help you to develop your understanding of these trends. (The professional development sources discussed in Chapter 2 might help.)
6. Based on what you have read in this chapter, how can you plan strategically for your future in technical communication?
7. Which skills have you developed that you believe will remain important in your technical communication career? Which new skills would you like to develop?

References

1 Allen, R. C. (2017). Lessons from the history of the future of work. *Nature, 550,* 321–324. https://doi.org/10.1038/550321a
2 Susskind, R., & Susskind, D. (2017). *The future of the professions: How technology will transform the work of human experts.* Oxford University Press.
3 Lasi, H., Fettke, P., Kemper, H. G., Feld, T., & Hoffmann, M. (2014). Industry 4.0. *Business & Information Systems Engineering, 6,* 239–242. https://doi.org/10.1007/s12599-014-0334-4
4 Information 4.0 Consortium. (n.d.). *About the Information 4.0 consortium.* https://information4zero.org/about/
5 Anthes, E. (2017). The shape of work to come: Three ways that the digital revolution is reshaping workforces around the world. *Nature, 550,* 316–319. https://doi.org/10.1038/550316a
6 Harari, Y. N. (2017). Reboot for the AI revolution. *Nature, 550,* 324–327. https://doi.org/10.1038/550324a
7 Følstad, A., & Skjuve, M. (2019, August). Chatbots for customer service: user experience and motivation. In *Proceedings of the 1st International Conference on Conversational User Interfaces.* ACM. https://doi.org/10.1145/3342775.3342784

8 Masycheff, A. (2018, May 22). Keeping technical writing jobs relevant with content automation technologies. *Intuillion*. https://intuillion.com/2018/05/22/keeping-technical-writing-jobs-relevant-with-the-content-automation-technology/

9 Lewis, S. C., Guzman, A. L., & Schmidt, T. R. (2019). Automation, journalism, and human-machine communication: Rethinking roles and relationships of humans and machines in news. *Digital Journalism, 7*(4), 409–427. https://doi.org/10.1080/216708 11.2019.1577147

10 Earley, S. (2018, January). AI, chatbots, and content, Oh my! *Intercom, 65*(1), 12–14.

11 Følstad, A., & Brandtzæg, P. B. (2017). Chatbots and the new world of HCI. *interactions, 24*(4), 38–42. https://doi.org/10.1145/3085558

12 World Economic Forum. (2020). *The future of jobs report 2020.* https://www.weforum.org/reports/the-future-of-jobs-report-2020/digest

13 Mehlenbacher, B. (2013). What is the future of technical communication? In J. Johnson-Eilola and S. A. Selber, *Solving Problems in Technical Communication* (pp. 187–208). University of Chicago Press.

14 Hoffman, A. (2018, January). The evolution of tech comm: Directing the content experience. *Intercom, 64*(1), 6–8.

15 European Commission. (n.d.). *EU data protection rules.* https://ec.europa.eu/info/law/law-topic/data-protection/eu-data-protection-rules_en

16 Acrolinx. (2020). *Content that feels like your strategy.* https://www.acrolinx.com/product/

17 Breuch, L. A. K. (2018). *Involving the audience: A rhetorical perspective on using social media to improve websites.* Routledge.

18 Upwork. (2020). *The world's work marketplace.* https://www.upwork.com/

19 Abel, S. (2020, January/February). It's time we start personalizing technical documentation experiences. *Intercom, 67*(1), 4–10.

20 Larson, M. S. (1977). *The rise of professionalism: A sociological analysis.* University of California Press.

21 Barnes, H. (2020). *The Covid-19 world and its impact on the future of work.* WorkDesign Magazine. https://www.workdesign.com/2020/03/the-covid-19-world-and-its-impact-on-the-future-of-work/

22 Friedman, Z. (2020, May 6). *How Covid-19 will change the future of work.* Forbes. https://www.forbes.com/sites/zackfriedman/2020/05/06/covid-19-future-of-work-coronavirus/#396e48b573b2

23 Deloitte Insights. (2020, May 15). *Returning to work in the future of work: Embracing purpose, potential, perspective and possibility during COVID-19.* https://www2.deloitte.com/us/en/insights/focus/human-capital-trends/2020/covid-19-and-the-future-of-work.html

24 Anderton, R., Jarvis, V., Labhard, V., Morgan, J., Petroulakis, F., & Vivian, L. (2020). *Virtually everywhere? Digitalisation and the EU area and EU economies* (Occasional Paper Series). European Central Bank. https://www.ecb.europa.eu/pub/pdf/scpops/ecb.op244~2acc4f0b4e.en.pdf

25 Johnson, T. (2020, May 26). Results of pandemic impact on tech comm survey. *I'd Rather Be Writing.* https://idratherbewriting.com/blog/results-of-pandemic-impact-on-tech-comm-survey/

26 Pratt, E. (Host). (2020, April 23). *Covid-19: The future for organisations and for technical communication* [Audio podcast]. Cherryleaf. https://cherryleaf.podbean.com/e/86-covid-19-the-future-for-organisations-and-for-technical-communication/

27 St. Amant, K. (n.d.). *Communicating about Covid-19.* https://communicating-about-covid19.weebly.com

28 Write the Docs Australia (2020, 28 May). *Tom Johnson and Kirk St. Amant: Remote discussion* [Video]. YouTube. https://youtu.be/RbQOFgBRnGk

29 Carliner, S. (2010). Computers and technical communication in the 21st century. In R. Spilka (Ed.), *Digital literacy for technical communication: 21st century theory and practice* (pp. 21–50). Routledge.

30 Wilson, G. (2001). Technical communication and late capitalism: Considering a post-modern technical communication pedagogy. *Journal of Business and Technical Communication, 15*(1), 72–99. https://doi.org/10.1177/105065190101500104

31 Rainey, K. T., Turner, R. K., & Dayton, D. (2005). Do curricula correspond to managerial expectations? Core competencies for technical communicators. *Technical Communication, 52*(3), 323–352.

32 Hoffmann, A. (2015, August 18). *Creating smart documentation and the future of technical writing.* The Content Wrangler. https://www.brighttalk.com/webcast/9273/166489/creating-smart-documentation-and-the-future-of-technical-writing

33 Rockley, A., & Gollner, J. (2011). An intelligent content strategy for the enterprise. *Bulletin of the American Society for Information Science and Technology, 37*(2), 33–39. https://doi.org/10.1002/bult.2011.1720370211

34 Baker, M. (2018, April 6). Time to move to multi-sourcing. *Every Page is Page One.* https://everypageispageone.com/2018/04/06/time-to-move-to-multi-sourcing/

35 OASIS. (2018, June 19). *Darwin Information Typing Architecture (DITA) Version 1.3 Part 3: All-inclusive edition.* http://docs.oasis-open.org/dita/dita/v1.3/dita-v1.3-part0-overview.html

36 Lanier, C. (2018). Toward understanding important workplace issues for technical communicators. *Technical Communication, 65*(1), 66–84.

37 OASIS. (2018, October 30). *Lightweight DITA: An Introduction Version 1.0.* http://docs.oasis-open.org/dita/LwDITA/v1.0/LwDITA-v1.0.html

38 Schema.org. (2020). *Welcome to schema.org.* https://schema.org/

39 Johnson, T. (2018, July 23). Adventures of a techie academic with Lightweight DITA (LwDITA): Conversation with Carlos Evia. *I'd Rather be Writing.* https://idratherbe-writing.com/2018/07/23/adventures-of-techie-academic-conversation-with-carlos-evia/

40 Loranger, H., & Nielsen, J. (2017, January 29). *Microcontent: A few small words have a mega impact in business.* Nielsen Norman Group. https://www.nngroup.com/articles/microcontent-how-to-write-headlines-page-titles-and-subject-lines/

41 Lebson, C. (2016). *The UX careers handbook.* CRC Press.

42 Johnson, T. (2020). 2020 developer documentation survey. *I'd Rather be Writing.* https://idratherbewriting.com/blog/developer-documentation-survey-2020/

Part III

Theories and Methods

The third part of the book is organized into three chapters that discuss the theories and methods upon which the book is based and that connect the more practical chapters to a research foundation. This section is important because it demonstrates that the study of technical communication in industry is connected to broader studies of professions and practice. It also explains how the book's content is grounded in research and data.

- Chapter 7 explains the theoretical perspectives that informed the book, focusing on theories of professionalization and practice and their applications in technical communication research.
- Chapter 8 explains the methods used to gather and analyze the data that informed the content of the first two parts of the book.
- Chapter 9 draws conclusions and proposes a strategic and research agenda for the technical communication profession and its practice.

Whether you are a student, teacher, or practitioner, these chapters help you to connect theory and research to technical communication practice in industry. They also encourage you to consider strategic directions for your work, your professional development, and your profession.

7 Theories of Professions and Practice

Introduction

All research studies have a theoretical and research foundation upon which they build. Because technical communication is a new field of work compared with many other professions, we might imagine that theories of professions and practice have less relevance for us. I believe that theories of, and research about, professions and practice help us to understand, explain, reflect on, and promote our work. They also help us to recognize ways in which our profession shapes our identity and we shape our profession. They help us to strategize, too, by enabling us to see what the profession has achieved and how it can develop in the future.

This chapter enables you to reflect on your profession and practice and to connect theory to practice in industry.

This chapter explores the two related theoretical and research bases of this book:

- Theories of **professions and professionalization** (from sociology) and academic studies about professionalization in technical communication. In the first section of the chapter, I explain my understanding of the theoretical foundation for professions generally and research about the profession of technical communication in particular.

- Theories of **practice** (from education and organizational studies research) and recent studies about aspects of technical communication practice. I explain my interpretation of theories of practice and discuss research into practice in technical communication.

- If you are a **technical communication student**, you will be able to recognize how professionalization and practice theories apply to your field of study.

- If you **work in industry as a technical communicator or** are **considering a career change into technical communication**, you will be able to use these theories to explore, explain, and analyze your profession and your practice.
- If you **are a technical communication teacher or researcher**, you can use this chapter as a teaching and study resource for undergraduate and graduate courses on theory in technical communication.

Profession and Practice: A Broad Perspective

This book's organization was guided by Pierre Bourdieu's theory of practice.[1] Bourdieu was a French sociologist, anthropologist, and philosopher whose work influenced many academic disciplines. He explained practice using the concepts of field, *habitus*, and capital. The first two parts of this book, about the profession (or field) and practice (*habitus*), separate out the broad professional perspective of what technical communication is from its everyday activities (how we work). This separation is an organizing principle for the book, but it is more than that, I hope. It offers a broader perspective than an analysis of either profession or practice alone could, and it enables you to think about technical communication beyond your current job or study program. At times, the separation of profession from practice is artificial. For the most part, though, its purpose is to help you to think about your field of work generally and your work patterns, material realities, and activities specifically.

Defining the Field and Habitus

Bourdieu used the analogy of a playing field, where the game to be played has a unique set of rules, the *habitus*.

The field is the **arena of activity**. Bourdieu and Wacquant call the field "a socially structured space in which agents struggle, depending on the position they occupy in that space, either to change or to preserve its boundaries and form" (p. 17).[2]

> A profession, for example, is a field with its own characteristics, in which workers are the agents striving to either improve or maintain the status quo.

The *habitus* comprises the activities we undertake in any field, our (often unconscious) **routine actions**, and how we adapt to situations. The *habitus* is the set of actions that make a field and sustain it.

> In a profession, the *habitus* can be defined as the activities that professionals do regularly and that fuse to become their practice.

Although *habitus*, for Bourdieu, represented primarily tacit knowledge, it can be learned, and he described this learning as "the long dialectical process, often described as 'vocation', through which the various fields provide themselves with agents equipped with the *habitus* needed to make them work" (p. 67).[3] Bourdieu also recognized that the borders between fields are sometimes poorly defined but nevertheless each field is recognizable by its unique rules. This is an interesting direction for us to consider in technical communication, a profession that shares boundaries with many others, as we saw in Chapter 1.

The Importance of "Capital"

Bourdieu[4] described various forms of "capital" that determine how class and power interact in different arenas of society. Some types of capital are tangible, like money, and others are less explicit, like class, prestige, and academic accomplishments. Within a profession, these types of capital are important:

- Cultural capital includes any form of academic or other qualifications or skills that are in demand. Cultural capital enables individuals to improve their social position.
- Symbolic capital consists of prestige markers, like prizes or titles. Being a member of some professions confers symbolic capital.
- Economic capital is the potential for remuneration afforded by a profession.
- Social capital is developed through networks and relationships.

These forms of capital help to explain why becoming a profession is a goal for many occupations. The more capital you have, the more powerful you are. Many studies about professions explore the impact of these and other forms of capital. Some professions, or fields, have more access to capital than others, as Driskill and Watts explain (p. 299):[5]

> The field determines what range of agency an individual may have and what cultural, economic, and social "capital" an individual may accumulate in order to advance to a higher position.

Just belonging to some professions gives you social and symbolic capital. For example, in Ireland in the mid-twentieth century, families considered it a great source of pride to have a priest, a physician, or, to a lesser extent, a teacher in the family. Being a relative of one of these professionals afforded social and symbolic capital.

In technical communication, many researchers and practitioners are concerned with our value to, and our status within, organizations. For these reasons, increasing our economic, social, cultural, and symbolic capital has been strategically important.

Professions and Professionalization

Professionalization theory examines how an occupation becomes a profession. This theory has been influential in technical communication: several studies have explored whether we are a profession and how we can become more professionalized.

What are Professions?

How is a profession distinguishable from a regular occupation? The term "profession" is often used loosely to describe any paid employment. The term "professional" is also contrasted with the term "amateur" to describe a person whose work "is devoted to an activity, as against one who is only transiently or provisionally so engaged" (p. 152).[6] In sociology, however, the term "profession" has a precise meaning. It describes a "field of work whose practitioners [have] gained control of their own training, admission to practice and evaluation of standards of performance" (p. 13).[7]

Because they have gained control of various aspects of training and performance, professionals have high status, public recognition, a monopoly to practice, a fixed educational path, codes of ethics, and a coherent career structure.[8,9,10,11] Until the 1970s, researchers described professions based on traits and on the relationship of the profession to society. This way of describing professions and other social structures is known as **structural functionalism**. The common traits are summarized in Box 7.1.

Box 7.1   Overview of professional traits

Professionals develop technical competence through a **recognized educational or training program**. The academic qualification gives them a **license to practice**, a clear **career path**, an **ethical framework** within which to operate, and **high status**. They **work to serve** rather than purely for personal profit, but the **high income** they earn is commensurate with the demand for the professional service and its perceived status. They are usually members of one or more **professional associations**. In the so-called "golden age"[12] of professions, in the mid-twentieth century, professionals were likely to be self-employed and therefore **in control of their practice and technique**.

These traits differentiate professionals from workers in regular occupations. They are an exclusive, elite group of workers (like physicians and lawyers), and they have a status and associated symbolic capital that are difficult to attain.

Technical communication is not strictly a profession if we compare its traits with those of established professions. It is an occupation offering "services based on expert knowledge but lacking the autonomy, service orientation, or prestige of traditional professions" (p. 280).[12]

I use the term "profession" to refer to technical communication throughout this book, however, to acknowledge that this is a distinct field of work with its own set of characteristics and workers who operate in a spirit of professionalism.

Although many researchers discussed professions uncritically and in generally positive terms, professions have negative traits, too. In the 1970s, several researchers began to examine the problematic aspects of professions. The negative traits they identified include monopolization, hegemony, elitism, and bureaucratization.[8,9] These negative traits have led to unfavorable public perceptions of some professionals (e.g., politicians).

Becoming a Profession

From the early 1900s to the present, studies have examined the nature and characteristics of professions and explored how occupations become professions. These studies help to explain how and why "occupations vary in the degree to which they are professionalised" (p. 93).[13] Occupations in the process of professionalization are known as minor professions,[14] or semi-professions.[15] Whereas occupations like law, medicine, and engineering are regarded as full professions, semi-professions (like social work) have only some traits of a profession.

Many occupations have sought to professionalize. For example, accounting and some health-care occupations have had relative success in professionalizing.[10] They have developed recognized academic and training programs and requirements, and practitioners cannot practice without accreditation. Others have had less success. For example, despite many academic and organizational attempts to professionalize journalism, journalists can still practice without regulation or a prescribed academic qualification. Technical communication, as a new profession, has not traditionally been regarded as a full profession, and this was a concern of much of our research in the 1990s and 2000s.

Technical Communication Professionalization

Professionalization projects are often spearheaded by professional associations and researchers. These projects can help us to understand what a profession is, examine our work, and decide how we want to professionalize. Steps to professionalize technical communication in the mid-twentieth century resulted in the establishment of professional associations, such as the Society for Technical Communication (STC), the Institute of Scientific and Technical Communicators (ISTC) in the UK, and tekom in Germany.

From the 1960s, academics and professional associations in technical communication worked to achieve professional status. In 1961, Israel Light, in an article entitled *Technical writing and professional status,*[16] asked: "What difference should it make to the technical writer whether he or she be considered a member of a 'trade;' a 'craft,' or a 'profession?'" His response indicated that the concern at the time was increasing our social, symbolic, and economic capital:

> What we are called makes a big difference. In the world of work and in the process of earning a living, the words 'profession' and 'professional' represent preferred social status, high income, and specialized competence.

Professionalization continued to be a big research focus until recently. Various researchers[17,18,19] explored the claims that technical communication could make to professional status, the benefits of professional status, and the steps needed for technical communication to become a profession. Identifying professional traits was a common way of analyzing occupations, including technical communication. Traits are a convenient yardstick to measure and describe a field of work and the strategic direction it needs to take. As Savage argued in relation to technical communication, "it seems likely that a mature profession must display most of the traits that historical and sociological studies suggest are typical of fully developed professions" (pp. 361–362).[17]

Publications about professionalization included a two-volume edited collection[20,21] and articles about projects like the Technical Communication Body of Knowledge (TCBOK).[22] In an influential 2004 article, Barbara Giammona[23] posed research questions about the future of technical communication:

- What is a technical communicator today?
- Which technologies are impacting on the field?
- What is the future role of technical communicators?
- How should we educate future practitioners?

Her research involved a survey of technical communication practitioners, academics, managers, and recognized leaders worldwide. She also interviewed experienced practitioners. At the time, she concluded that technical communication was an important field of work with a clear purpose that was changing rapidly and needed a clearer strategic direction. Although her article is now almost two decades old, the questions and conclusions still seem important.

By the time of the publication of a two-part special issue of the *Technical Communication* journal on professionalization in 2011/2012, the guest editor, Nancy Coppola, found from many initiatives "an appetite for professionalization [...] a growing collective consciousness, but certainly not collective agreement, for professionalization" (p. 283).[24]

Since that time, occasional articles about the professionalization of technical communication appear (see, for example, a 2020 study of legitimacy and professionalization based on interviews with practitioners[25]), but the heat seems to have left the debate.

Problems with the Appeal to Professionalization for Technical Communication

In my research, I was also exploring professional concerns (I explain this background more in Chapter 8). In an article in the 2012 *Technical Communication* special issue,[26] I examined the blogs of prominent technical communication industry practitioners regarding their views about professionalization topics. I concluded that although these practitioners wrote about professional issues, they were not concerned explicitly with professionalization. In a later article,[27] I described the findings of surveys and interviews about professional topics with practitioners based in Ireland. Although the participants in that study were keen to be more involved in a professional community, there was no obvious professionalization agenda among technical communicators in Ireland, for many reasons. Most of the respondents worked in corporations, and common traits (like autonomy, market closure, and a prescribed educational path) that are the hallmark of traditional professions were irrelevant to them. Corporations control how they recruit employees, and as long as applicants show they can do the job, academic qualifications may not be required. Employers are often satisfied with relevant experience instead; indeed, many technical communicators have not chosen to study technical communication in a formal way. Furthermore, corporate workers have limited autonomy, and managers often control work practices and projects.

Because technical communicators work in diverse roles, with a variety of job titles, in a range of industries, in teams, as lone writers or as freelancers, I concluded that seeking to be recognized as a profession was not an appropriate goal. I remained convinced, however, of the strategic benefit to us academics and industry practitioners, of deepening our professionalism, by understanding and strengthening our profession. This conviction led me to explore alternative approaches to studying the profession.

Alternative Approaches to Studying Occupations

For many reasons, research into the traditional professions and their characteristics has stalled since the "golden age" of the mid-twentieth century.[12] Research about work organization and practice generally, and within expert disciplines specifically, has persisted, however. Some of the concerns of the research about professions, including research into professional development, professional associations, and professional practice, also remain important and active.[28] Current professional studies focus on "processes and the social actors who move them forward—individual workers, employing organizations, and formal and informal occupational groups" (p. 291).[12] These concerns, as you will have seen throughout the book, became my focus.

The Symbolic Interactionist Perspective

Most researchers have moved on from drawing "a hard and fast line between professions and occupations" (p. 780).[29] We see more variety in approaches to studying occupations and labor, as structural functionalism has become less prominent in sociological research.[10] This move has incorporated a pivot toward symbolic interactionist research. Symbolic interactionism examines the impacts of relationships on individuals in society. Blumer described it as both a theory and a method, and he identified core principles (p. 50):[30]

- Individually and collectively, people construct meaning.
- People make sense of their worlds through their interactions with others.
- Meaning changes over time. The "complex interlinkages ... that comprise organizations, institutions, divisions of labor ... are not static affairs" (p. 50).

This perspective enables us to study and think about professions in ways that are more human-centered. Instead of describing the traits of a particular type of work, we explore our actions, our interactions, how we explain our worlds to others, and how we understand our careers.[10]

In technical communication, a symbolic interactionist perspective enables us, as researchers, teachers, students, and practitioners, to examine our experiences within the profession and our interactions with the profession, with the practice, with one another, and with individuals in other professions.

In an evolving field and working environment like technical communication, understanding and meaning are not static, and individuals construct meaning all the time. Carolyn Miller (p. 615)[31] explained the communal and community interactions in our development of knowledge and understanding:

> [W]hatever we know of reality is created by individual action and by communal assent. Reality cannot be separated from our knowledge of it; knowledge cannot be separated from the knower; the knower cannot be separated from a community.

Symbolic interactionism examines individual and collective actions within an occupation and how these actions alter its status. It incorporates concepts of professionalism and professional identity. This is the perspective I have adopted in this book.

Professionalism and Professional Identity

Professionalization was problematic because it encouraged a focus on traits of an occupation while disregarding individual practitioners, individual differences among practitioners, and the values of the profession.[29] It de-emphasized professional*ism* and professional service. Re-orienting to professionalism enables us to view the profession as a field and our own activities as the practice within that field (or the *habitus*, to use Bourdieu's term). The profession is a site where we perform activities rather than the reason that we perform them. Professionalism leads us to consider our professional identities (discussed in more detail in Chapter 3). Your perception of yourself is affected by your occupation and by your interactions within your occupation. For some individuals, and in some professions, that identification is stronger. Magali Larson (p. 228, original emphases)[9] described how our personal and professional identities become entwined:

> What *typically* binds a professional to his profession, and therefore to its elites, has to do *with the character of work itself.* [...] *All* occupations which involve special skills and special worlds of work shape to some extent the worker's personality of self-presentation; initiation into techniques, languages or jargons, ways of dressing, and mannerisms, identify the individual with his occupation for himself and for outsiders. The pleasure of 'talking shop' is not restricted to the professions, nor is the anticipatory socialization which prepares an individual to look and act like people in his chosen field are supposed to. [...] the result of *all* occupational socialisation is the same; it tends to create conformity and to identify people with work roles, and also with the stereotypes of those roles that are held in the larger society.

As this excerpt suggests, we all identify with our work to some degree. Over-identification can become stifling, however. When we identify very strongly with our profession, we may act conservatively and conform to our perceived societal roles.

Professional Identity and Performance

Erving Goffman[32] described professional identity as performance. We take on roles or have them thrust upon us, whether at work or in our personal lives. Goffman suggested that our interactions, how we present ourselves, and how we perceive others are inherently performative. Professionals learn how to perform from other professionals, and we also learn based on the role we must perform. Our performance may seesaw between sincere and cynical through the course of our careers.

Goffman believed that work that can be dramatized effectively is perceived by the public to be more valuable. He used the examples of police work and surgery to illustrate how some roles are more easily dramatized. The number of films and television shows set in hospitals and police precincts bears out this reasoning. He extrapolated that it may be more difficult to justify the costs associated with, and perhaps more difficult to explain, or gain respect for, work that does not lend itself to dramatization. Goffman used the example of the "invisible" work that goes into preparing a short radio talk.

> Technical communication may be poorly understood in part because it is not easily "dramatized." Indeed, we do not see many technical communicators in the public domain, on television, in films, or in public discourse, even when our contributions could be highly relevant.

The metaphor of performance can be extended. We can view our workplaces as stage settings and our work activities as a dramatization, the play we are performing. This perspective maps to Bourdieu's separation of field and *habitus*. This idea of work as performance on a stage (our profession) interested me and influenced the study for this book. The practice narratives are personal accounts by industry practitioners who expressed their individual experiences and activities in narrative forms that enabled them to interpret and construct their own stories.

Combining our Professional Identities

Identity is complicated. Professionals may have different social identities in professional and personal capacities. Goffman (p. 58)[32] explained:

> [A] professional man may be very willing to take a modest role in the street, in a shop, or in his home, but, in the social sphere which encompasses his display of professional competency, he will be much concerned to make an effective showing. In mobilising his behaviour to make a showing, he will be concerned not so much with the full round of the different routines he performs but only with the one from which his occupational reputation derives.

According to Goffman, some aspects of our professional identity bleed into our personal lives but others do not. Sometimes, you will experience tension between the various identities you inhabit in different situations. Even within a single occupation or organization, you may behave differently in different settings, and these differences contribute to your composite professional identity. For example, you may be more assertive in one team than in another. If you take a leadership role in a team or organization, aspects of your identity will come to the fore that were not prominent previously. In Chapter 3, we explored how our professional identity is also shaped by community interactions at work. Your interactions within many different communities shape your combined professional identity (e.g., your interactions on multiple teams or committees, your personal and professional communities, and your local and global communities).

Factors that may Dilute Professional Identity

In the past, especially in traditional professions, individuals identified very strongly with their profession. If you were a physician, that was a big part of who you were, even outside of work. In many professions, individuals now have a weaker sense of professional identity. Recent studies have suggested some reasons why we may not identify as strongly with our profession as individuals did in previous generations.

- **Diversity** in training, social backgrounds, organizational cultures, and roles within an organization may lead to a weaker sense of professional identity.[12] For example, in technical communication, the increasing diversification in job titles and training that we discussed in Chapters 1 and 2 may lead you to identify less with your profession. You might not identify with the "technical communication" label if you have a different study program or job title.
- If you work mostly in **cross-functional teams** with colleagues from many different disciplines, as is often the case in technical communication, over time you may identify more strongly with your team than with your profession. Indeed, in order to be successful, you may need to identify with the team, although this identification can coexist with your professional identity.[33]
- In large organizations with **dominant organizational cultures**, some employees identify more with their organization than with their profession. For example, a few years ago I interviewed technical communicators who all worked in one organization. They belonged to a large publications team that had several resources for writers. They identified very strongly with the organization. They used corporate terminology, depended on the corporate style guide, and were all members of a local publications group. None of them was a member of a professional association, however, and they did not network with technical communicators in other organizations.
- As we saw in Chapter 3, **professional associations have less influence** than in the past, as many practitioners join free online communities instead. Professional associations are essential supports for professionalism and professional identity; their diminishing influence may in turn reduce the influence of practitioners and the perceived value of their practice.

Outcomes of Professionalization for Technical Communication

According to the structural functionalist (or traits) model, a successful professionalization project would result in regulated training, professional bodies, license to practice, and the development of a research base. In technical communication, we have achieved some, though not all, of these outcomes. The many projects and studies, and the attention given to professionalizing, have resulted in several positive initiatives, however.

1. **Emphasis on professional identity**
    To empower users of technology, the workers who explain technology also need to be empowered to engage with their jobs and to participate in their profession. Part of the empowerment process involves enabling technical communicators to identify with their profession, to recognize themselves as professionals, and to contribute to, acknowledge, and explore the boundaries of their practice. This is an ongoing project in technical communication.

2. **Education and training programs**
    Chapter 2 provides an overview of the current state of technical communication education and training globally. Of course, more work is needed; no education programs are available in this profession in many regions of the world. Nevertheless, the number of programs available at all levels demonstrates that the steps taken toward professionalization have had very positive outcomes for education.

3. **Projects and resources**
    Chapter 2 also describes two projects that identified technical communication competencies and that have led to increased awareness of the skill sets and activities in this profession: the STC-supported TCBOK and the tekom Europe-supported and European Union-funded TecCOMFrame project. These are two of many projects led by professional associations that increase the status, recognizability, and stability of this profession.

4. **Strong communities**
    Although professional associations struggle to maintain membership numbers, many communities support technical communicators, as outlined in Chapter 3. Supportive communities enable us to learn, develop our skills, solve problems, and network. They also enable us to contribute and to further the goals of the profession.

5. **Research**
    Although there is a divide between academia and practice, the move to professionalize led to research projects that contributed to the profession and that examined professional identity. Furthermore, as technical communication has matured, it has become less important to define ourselves, and new avenues for research consistent with a more mature discipline have also become prominent. More published studies have explored aspects of technical communication practice, coinciding with a (re)turn to practice and a broader movement to theorize interactions and practices, especially in work and organization studies research.

The Practice Turn in Theory and Research

To understand any occupation, especially new occupations like technical communication, we need to explore not only the features of the profession but also its practice(s). The past two decades have seen an increasing research emphasis on practice in many disciplines. The communities-of-practice model, popularized by Wenger,[34] is an influential, but certainly not the only, practice theory. Many recent works have examined how to theorize and study practices.[35,36,37,38]

What is Practice?

Practice is studied in multiple disciplines (e.g., philosophy, sociology, anthropology, cultural theory, history, science, and technology).[36] Although Bourdieu never defined practice explicitly, it was a central concern of many of his works, and his separation of field and *habitus* corresponds to my dual emphasis on profession and practice in this book. Over the past two decades, several researchers have offered definitions of practice:

- Schatzki defined practice broadly (p. 2):[36] "Most thinkers who theorize practices conceive of them, minimally, as arrays of activity." In later work, he summarized various practice theorists to describe a practice as "an organised constellation of different people's activities [...] Practices are grounded in organised human activity" (p. 13).[37]
- Hui saw practices "as patterns created through the bringing together of a set of activities, materials, understandings and skills" (p. 52).[39]
- Nicolini (p. 3)[35] suggested that "all practice theories foreground the importance of activity, performance, and work in the creation and perpetuation of all aspects of social life."

> The commonalities of these definitions that are relevant to this book are **organized patterns of human activities, materials, and skills in a given arena (in this case, a work arena).**

Communities of Practice

The communities-of-practice framework was first proposed by Lave and Wenger[40] and was extended by Wenger.[34] Wenger examined the concept of practice from a community perspective: what it means to do a particular job and to identify with one's job and, especially, how all practice is social. The community-of-practice model has four interdependent components: community, practice, identity, and meaning. Communities of practice foster both community and professional identity. Even when we appear to work individually, we understand our work through the prism of our interactions with others. We learn together how to negotiate the boundaries of what we are required to do, implicitly and explicitly, in a work setting: our practice.[34] We develop this understanding and learn our practice through socialization and education (both formal and informal). Social interactions such as shared discourse, information flow, shared ways of doing things, and collaboration help us to understand our practice. "The idea of a community of practice is that people typically come together in groupings to carry out activities in everyday life, in the workplace and in education" (pp. 1–2).[41] All practice involves social and community participation.[42]

A profession might be regarded as a *constellation* rather than a community of practice.[34] Wenger outlined common characteristics of a constellation of practice, including shared roots, related activities, shared artifacts, "overlapping styles or discourses," and common membership. In relation to technical communication, this set of commonalities encompasses how the profession has developed and its core competencies, tools, genres, and communities.

Communities and constellations of practice have many benefits:

- They enable members to feel a sense of belonging and to identify with their work.
- They also help us to make sense of our work. "We all have our own theories and ways of understanding the world, and our communities of practice are places where we develop, negotiate, and share them" (p. 48).[34]
- Despite constant change in many types of workplace and profession, both the community or constellation and the organization provide continuity and a sense of stability, and most practices balance on a thread of tension between continuity and instability.

Communities of practice are important in a field like technical communication with "a diversity or dichotomy of professionals" (p. 24).[43] Within the "complex and dynamic environment" of technical communication, "'community' is emerging as a critical element both in supporting individual professional development and in developing the future of the profession itself" (p. 277).[44] Chapter 3 describes many online and face-to-face communities of practice that support technical communication practitioners. These communities help us to understand, to engage with, and to contribute to our profession and to develop our practice.

The concept of **legitimate peripheral participation** explains how our status gradually evolves in practice as we progress from newcomer to expert. When we begin to practice in any field, we learn by observing and through "peripheral participation." The time taken to learn from others (e.g., through observing or being mentored) before undertaking big responsibilities gives us confidence. In Chapter 2, several practice narratives described how their work and learning histories exemplify legitimate peripheral participation.

Variation in Practice

Throughout this book, we have explored prominent themes of variety and diverse work practices in technical communication. Variation is normal and is a feature of all practices,[39] not only of technical communication. Even within well-established practices, everyone will have a slightly different method of working. Culture and time also lead to variations in how practices are enacted (e.g., funeral rituals vary substantially from one culture to another and can change over time).[39] Variation is also essential. It helps the practice to grow and ensures it is sustainable. Wenger explained: "because the world is in flux and conditions always change, any practice must constantly be reinvented, even as it remains 'the same practice'" (p. 94).[34]

Of course, there is a tension between variety and stability. Variations must have limits. Too much variety means that there is no core to a practice, and the lack of distinctiveness makes the practice unsustainable. A level of "tolerable variation" (p. 56) can exist, beyond which the practice is no longer recognizable.[39] We base our understanding of practices on a general idea of what we believe the practice ought to be, our "normative conceptions" (p. 2).[42] Therefore, when studying a practice, we need to explore what "normative" means for that practice. What is a typical set of activities? In some practices, the set of activities is relatively finite and stable. That is especially true when practitioners specialize. For example, many legal professionals specialize in a facet of law practice, where the range of activities is predictable. Even within such well-established practices, however, practitioners' methods and activities vary. Variations in practice gradually become normalized through experience, tacit knowledge, and the individual situation. The organization and individuals within the community develop mechanisms, such as policies or forms, to foster a sense of stability.

Researching Practice

Practice studies take many forms, depending on their objectives. The practice approach to research is "demarcated as all analyses that 1) develop an account of practices, either the field of practices, or some subdomain thereof [...] or 2) treat the field of practices as the place to study the nature and transformation of their subject matter" (p. 11).[35]

> This book takes both of those approaches: it examines the profession of technical communication (the field of practice, or the profession) and also explores contemporary technical communication practice(s).

Qualitative methods help us to understand how practice is carried out.[42] Nicolini presented a toolkit of approaches to studying practice with research questions and supporting theories.[35] In this book, I took up one of his suggestions: to "zoom in" to the detail of practices and "zoom out" to regard the bigger picture and the relationships to other practices. Practice theorists also adopt aspects of the symbolic interactionist approach. For example, Gherardi defined work as "being-in-the-world tied to the accomplishment of a project through physical activities that are situated in space and time" (p. 7).[42]

Practice studies help to explain practice to practitioners maybe as much as they help students and researchers. Gherardi explained how "practices are as opaque to practitioners as to researchers, but precisely for this reason, their description and reflection are potential means to empower practitioners" (p. 5).[42]

I considered this to be an important reason to study industry practice in technical communication and to include practitioners as both a primary data source and a primary target audience. Practitioners have as much to learn from one another as students, teachers, and researchers have to learn from the collective narrative contributions. In Chapter 8, I describe the ways I conducted research for this book, the steps I took, and the methods I used and their purposes.

Practice Studies in Technical Communication

Practice studies are appropriate for fields that have to adapt to change. The practice "idiom" can "resonate with the contemporary experience that our world is increasingly in flux and interconnected, a world [...] in which boundaries around social entities are increasingly difficult to draw" (p. 2).[35] Several practice-oriented articles and books have been published about technical communication in the past two decades. Although many of these studies do not discuss practice theories explicitly, they correspond to the types of studies that theorists like Schatzki, Nicolini, and Gherardi recommend.

Variation in Technical Communication Practice

The boundaries of many newer fields of work, including technical communication, are indistinct, and disciplines themselves are constantly adapting to new technologies and work practices. We know from research, from our own experiences, and from reading about other technical communicators' experiences that variation is an essential characteristic of our practice. We have diverse job titles and diverse roles. Sometimes, it is difficult to recognize who we are because of the variations in practice, but these variations, paradoxically, are also part of our identity. Given that variation and unclear boundaries are complex realities for technical communicators, it is important to study the practices and activities that unify us, too. It helps us to understand the "normative" conceptions of technical communication.

Many studies of technical communication practice have been concerned with this theme. Between 2000 and 2010, Rachel Spilka wrote several articles and chapters that described disciplinary ambiguity in technical communication. She called this a "profound outcome" of digitalization (p. 5).[45] She listed several indicators of this disciplinary ambiguity, including continual and significant decreases in the membership of professional associations like STC, the popularity of cross-functional teams, and changing job titles and fields, leading to a situation where technical communicators identify as multidisciplinary.

Practice Narratives in Technical Communication

Because of my research interest in narratives of practice, I explored existing narratives of technical communication. Technical communicators do not often surface in popular culture, but here are a couple of examples:

Figure 7.1 Tina the Technical Writer. DILBERT © 2012 Scott Adams, Inc. Used By permission of ANDREWS MCMEEL SYNDICATION. All rights reserved.

- Robert M. Pirsig's narrator in the 1974 philosophical novel *Zen and the Art of Motorcycle Maintenance* was an early public face of technical writing. Pirsig's narrator edited technical manuals for a living, and much of the novel is concerned with articulating a philosophy of quality, technology, people's attitudes to technology, and language to describe technology.[46]
- The Dilbert cartoon series features Tina the Technical Writer. Tina's persona is a maligned martyr who struggles to have her work respected. Figure 7.1 shows a common scenario for Tina.

Some academic work has explored narratives of technical communication practice. In 2001, Savage and Sullivan[47] published a collection of narratives of technical communicators' experiences of starting out in their work, technical communication practices, and home and work lives. Although many of the narratives described roles writing about software and other information technology (IT) applications, other types of technical writing work (e.g., medical writing, copywriting, and engineering documentation) also featured. The contributors worked in corporations, universities, and their own homes. Some of these narratives were auto-fictional, whereas others described personal experiences. This collection provided authentic insights into the variety of technical communication roles and opportunities and also into technical communicators as individuals, with personalities, families, personal histories, and interests outside of work. By humanizing a relatively obscure practice in this way, it encouraged practitioners to explore their professional agency and demonstrated how in technical communication, as in other professions, "personal life and professional life can become intertwined" (p. xxvi). Its perspective was limited, however, because most contributions were from writers living and working in North America. It is also somewhat dated now, almost 20 years later.

More recent academic work has explored narratives of technical communication program administrators.[48] Other sources that provide contemporary narratives of the technical communication profession include the following:

- Several YouTube clips describe days in the lives of technical communicators. One example by a technical writer at Wizeline, a software development company, has over 25,000 views.[49]

- The ISTC has a regular series in its *Communicator* magazine, "A Day in the Life of a Technical Communicator." The UK-based contributors describe typical workdays.[50]
- *Intercom*, the magazine of the STC, featured a "My Job" segment in several issues. These one-page features provide insight into a variety of technical communication roles.
- Several bloggers have written posts about their typical days. Tom Johnson[51] characterized the role as constant task management. A "Technical Writing is Easy" blog post summarized "typical" days of a lone writer, a team leader, and a writer using a Docs-as-Code methodology.[52]

Stories of technical communicators at work give us insight into real-life experiences and workplaces. Gathering personal accounts of industry practitioners' experiences also enables them to express their individual experience and helps them to think about their role, identity, and contributions.

Research Studies of Workplaces and Practices

Studies of workplaces and practices are a prevalent research theme. Hart-Davidson[53] described three essential work patterns of technical communication practice: information design, user advocacy, and writing stewardship. His examination exposed the broad potential scope of technical communication practice. By contrast, many workplace studies are localized and have a specific theme. Here are some examples:

- Brady[54] explored communication practices of female professional communicators.
- Edenfield[55] and Colton and Holmes[56] studied democratic workplaces such as cooperatives.
- Petersen[57] reported on the findings of interviews with 39 US-based women practitioners. She concluded that the role is misunderstood and undervalued, and she urged practitioners to take steps to articulate their value and program directors/department heads to develop student collaborations with other disciplines.
- Brumberger and Lauer[58] developed personas of practitioners on the basis of data from an analysis of job postings, a survey, interviews, and "embedded observations" of technical communicators.

Many newer professions could be described as postmodern, especially when they are vulnerable to the fluctuations of corporate and globalized work patterns and technologies. Technical communication workplaces have been characterized as postmodern[59,60] and post-postmodern[61] because of the unpredictability in this profession. Lanier[62] discussed the changing workplace and its impact on perceptions of the job. Although he did not use the term "postmodern," his conclusion echoed the fundamental principle that constant change is the new reality in technical communication.

Although our work patterns may be varied, research in North America has exposed a lack of diversity among technical communication practitioners. Over 80% of respondents to the 2018 STC census identified as White.[63] When most practitioners have similar ethnic and cultural backgrounds, we are less likely to consider diverse users, and we may be unaware of our unconscious biases. We need to examine why our profession lacks diversity. The social justice turn (or shift in focus) in research helps us to think about our professional identity and our professional obligations through the lenses of diversity, fairness, equality, and inclusion. In the past decade, research about social justice in technical communication has explored how identity markers (like race, gender, and ability) impact the experiences of information users.[64]

In tandem with examining social justice, we also need to consider global identities and experiences, an area of limited research focus to date. Although many studies about technical communication practice have a North American focus, some studies[27,65,66] have also explored international practice concerns. Brumberger and Lauer[67,68,69] used data from job advertisements to catalogue technical communication activities, primarily in the US but also in the UK, Ireland, and India.

Although research about practice has been active, it is somewhat disjointed. At the same time, research about practice with an international focus is significantly under-represented in technical communication research.

Education and Practice Intersections

It is important for educators to understand changing practice and to consider how these changes may be reflected in an academic curriculum. Several recent studies explore the intersections of education and practice:

- Watts[70] analyzed how research projects based in technical communication workplaces could increase students' and teachers' understanding of technical communication practice.
- Wilson and Wolford[61] examined how to prepare students to work in postmodern and post-postmodern workplaces.
- Lanier[62] explored workplace trends and practices and revealed how they could map to education programs.
- Stanton[71] surveyed hiring managers and analyzed job advertisements to determine entry-level technical and professional communication (TPC) competencies. She compared her data with the findings from a 2013 study[72] of undergraduate TPC programs in the US. Her analysis showed that academic programs do a good job of preparing students for entry-level positions.

Studies have also investigated the links between research and practice and have found a pronounced gap. They demonstrate that our research is not keeping abreast of changing practice, leading to a profound gap between research and practice,[73] a homogenous research landscape that does not include industry voices,[74] and a lack of coherence and strategic direction in our research.[75] Rude noted how, because of "its expansiveness, [our] research has also largely turned away from practice, missing an opportunity to enhance the practice of professional communicators and to build the connections between the academic and practice parts of the field" (p. 370).[76] It is worth noting that the divide between academia and practice is pronounced in many fields.

Gherardi[42] concluded her volume on practice-based studies by encouraging readers to conduct action research that "employs the representation of practices as a means to empower practitioners" (p. 210). Her work spurred me to attempt to represent technical communication practices, gather input from practitioners, and potentially empower practitioners through their involvement. I am also convinced that we need "bigger picture" research and more and stronger ties between academia and practice, something that many of us consider necessary but that we have moved away from in our research.

> To achieve a change, the core problem needs to be clarified. When the issue is complex, the solution needs to consider the bigger picture and not just the parts. Telling stories from everyday practice and to systematically reflect and analyse those in interprofessional groups can create opportunities for enhanced understanding as well as be a vehicle for future change of practice.

This quote is from a study (p. 548)[77] that explored how to plan hospital discharges, apparently quite a distant topic from technical communication. For me, though, it rang true for my intentions in this study: to explore the "bigger picture" by involving practitioners in reflecting on their profession and practice and with a long-term potential for change. I conducted the research for this book to take a step toward exploring the "bigger picture". In the next chapter, I describe how I conducted the study.

Summary of Chapter 7

This chapter explores the theories that informed my research for, and organization of, this book. I adopted two complementary approaches: professions and practice.

- The organization of the book reflects Bourdieu's theory of practice and his separation of the concepts of field (profession) and *habitus* (practice).
- In sociology, professions are occupations that have high levels both of autonomy and of regulation. According to professionalization theory, occupations differ in the degree to which they are professionalized.
- The research agendas in many fields, including technical communication, have at critical times prioritized professionalization as a process and a goal, often using traits to demonstrate the extent to which an occupational field is professionalizing.

- Research into professions has stagnated in the past few decades, however, coinciding with an increase in practice-focused research, known as the practice turn.
- Practices are patterns of human activity in a particular arena. The focus in this book is on work practices.
- Although all practices exhibit variety, we also need to be able to identify the norms of a practice in order to recognize it.
- The communities-of-practice framework enables us to consider practice by focusing on learning, meaning, and identity. This framework explains how our understanding of practice, and the practice itself, is shaped through social interactions in communities.
- In technical communication, several recent studies discuss aspects of practice. These studies make strong individual contributions to our understanding of technical communication practice.
- Other narratives of technical communication practice provide additional insight into possible work and careers and help you to identify with the profession.
- Nevertheless, we currently lack a "bigger picture" research or strategic agenda that would help us to develop our profession and its practice. That "bigger picture" is the starting point of this book.

Discussion Questions

1. Find an example of a research article about professional characteristics or professionalization in technical communication. What are the article's key themes? What research directions does the article recommend?
2. Find an example of a research article about an aspect of technical communication practice. What are the article's key themes? What research directions does the article recommend?
3. Based on your understanding of this chapter, identify intersections between professionalization and practice.
4. What have you learned from this chapter that influences how you think about your practice, your work, and your career?

References

1 Bourdieu, P. (2010). *Outline of a theory of practice* (R. Nice, Trans). Cambridge, UK: Cambridge University Press. (Original work published 1972)
2 Bourdieu, P., & Wacquant, L. J. D. (2007). *An introduction to reflexive sociology.* Cambridge, UK: Polity Press. (Original work published 1992)
3 Bourdieu, P. (2010). *The logic of practice* (R. Nice, Trans.). Polity Press. (Original work published 1980)
4 Bourdieu, P. (2010). *Distinction* (R. Nice, Trans). Routledge. (Original work published 1970)
5 Driskill, L. P., & Watts, J. Z. (2010). Integrating experts and non-experts in mathematical sciences research teams: A qualitative approach. In J. Conklin & G. F. Hayhoe (Eds.). *Qualitative research in technical communication* (pp. 308–344). Routledge.

6 Flexner, A. (2001). Is social work a profession? *Research on Social Work Practice, 11*(2), 152–165. (Original work published in 1915)

7 Collins, R. (1990). Changing conceptions in the sociology of the professions. In R. Torstendahl & M. Burrage (Eds.), *The formation of professions: Knowledge, state and strategy* (pp. 11–23). London: Sage.

8 Johnson, T. J. (1972). *Professions and power.* Macmillan.

9 Larson, M. S. (1977). *The rise of professionalism: A sociological analysis.* University of California Press.

10 MacDonald, K. M. (1995). *The sociology of the professions.* Sage.

11 Freidson, E. (1999). Theory of professionalism: Method and substance. *International Review of Sociology, 9*(1), 117–129. https://doi.org/10.1080/03906701.1999.9971301

12 Gorman, E. H., & Sandefur, R. L. (2011). "Golden age," quiescence, and revival: how the sociology of professions became the study of knowledge-based work. *Work and Occupations, 38*(3), 275–302. https://doi.org/10.1177/0730888411417565

13 Hall, R. H. (1968). Professionalization and bureaucratization. *American Sociological Review, 33*(1), 92–104. https://doi.org/10.2307/2092242

14 Glazer, N. (1974). The schools of the minor professions. *Minerva, 12*(3), 346–364. https://doi.org/10.1007/BF01102529

15 Krejsler, J. (2005). Professions and their Identities: How to explore professional development among (semi-) professions. *Scandinavian Journal of Educational Research, 49*(4), 335–357. https://doi.org/10.1080/00313830500202850

16 Light, I. (1961). Technical writing and professional status. *Journal of Chemical Documentation, 1*(3), 4–10.

17 Savage, G. J. (1999). The process and prospects for professionalizing technical communication. *Journal of Technical Writing and Communication, 29*(4), 355–381. https://doi.org/10.2190/7GFX-A5PC-5P7R-9LHX

18 Spilka, R. (2002). Becoming a profession. In B. Mirel & R. Spilka (Eds.), *Reshaping technical communication: New directions and challenges for the 21st century* (pp. 97–110). Lawrence Erlbaum Associates.

19 Savage, G. J. (2003). Introduction: towards professional status in technical communication. In T. Kynell-Hunt & G. J. Savage (Eds.), *Power and legitimacy in technical communication: The historical and contemporary struggle for professional status* (pp. 1–12). Baywood.

20 Kynell-Hunt, T., & Savage, G. J. (Eds.). (2003). *Power and legitimacy in technical communication (Volume I): The historical and contemporary struggle for professional status.* Baywood.

21 Kynell-Hunt, T., & Savage, G. J. (Eds.). (2004). *Power and legitimacy in technical communication (Volume II): Strategies for professional status.* Baywood.

22 Coppola, N. W. (2010). The technical communication body of knowledge initiative: An academic-practitioner partnership. *Technical Communication, 57*(1), 11–25.

23 Giammona, B. (2004). The future of technical communication: How innovation, technology, information management, and other forces are shaping the future of the profession. *Technical Communication, 51*(3), 349–366.

24 Coppola, N. W. (2011). Professionalization of technical communication: Zeitgeist for our age introduction to this special issue (part 1). *Technical Communication, 58*(4), 277–284.

25 Rosselot-Merritt, J. (2020). Fertile grounds: What interview of working professionals can tell us about perceptions of technical communication and the viability of technical communication as a field. *Technical Communication, 67*(1), 38–62.

26 Cleary, Y. (2012). Discussions about the technical communication profession: Perspectives from the blogosphere. *Technical Communication, 59*(1), 8–28.

27 Cleary, Y. (2016). Community of practice and professionalization perspectives on technical communication in Ireland. *IEEE Transactions on Professional Communication, 59*(2), 126–139. https://doi.org/10.1109/TPC.2016.2561138

28 Adams, T. L. (2015). Sociology of professions: International divergences and research directions. *Work, Employment and Society, 29*(1), 154–165. https://doi.org/10.1177/0950017014523467

29 Evetts, J. (2013). Professionalism: Value and ideology. *Current Sociology, 61*(5–6), 778–796. https://doi.org/10.1177/0011392113479316

30 Blumer, H. (1969). *Symbolic interactionism: Perspective and method.* Prentice-Hall.

31 Miller, C. R. (1989). What's practical about technical writing? In B. E. Fearing & W. K. Sparrow (Eds.), *Technical writing: Theory and practice* (pp. 14–24). MLA.

32 Goffman, E. (1959). *The presentation of self in everyday life.* Doubleday.

33 Mitchell, R. J., Parker, V., & Giles, M. (2011). When do interprofessional teams succeed? Investigating the modifying roles of team and professional identity in interprofessional effectiveness. *Human Relations, 64*(10), 1321–1343. https://doi.org/10.1177/0018726711416872

34 Wenger, E. (1998). *Communities of practice: Learning, meaning, and identity.* Cambridge University Press.

35 Nicolini, D. (2012). *Practice theory, work, and organization: An introduction.* Oxford, UK: Oxford University Press.

36 Schatzki, T. R. (2001) Introduction: Practice theory. In T. R. Schatzki, K. K. Cetina & E. von Savigny (Eds.), *The practice turn in contemporary theory* (1–14). Routledge.

37 Schatzki, T. R. (2012). A primer on practices: Theory and research. In J. Higgs, R. Barnett, S. Billett, M. Hutchings & F. Trede (Eds.) *Practice-based education: Perspectives and strategies* (13–26). Brill/Sense.

38 Hui, A., Schatzki, T., & Shove, E. (2017). *The nexus of practices: Connections, constellations, practitioners.* Routledge.

39 Hui, A. (2017). Variation and the intersection of practices. In A. Hui, T. Schatzki & E. Shove (Eds.), *The nexus of practices: Connections, constellations, practitioners* (pp. 52–67). Routledge.

40 Lave, J., & Wenger, E. (1991). *Situated learning: Legitimate peripheral participation.* Cambridge University Press.

41 Barton, D., & Tusting, K. (Eds.). (2005). *Beyond communities of practice: Language power and social context.* Cambridge University Press.

42 Gherardi, S. (2012). *How to conduct a practice-based study.* Edward Elgar.

43 Kline, J., & Barker, T. (2012). Negotiating professional consciousness in technical communication: A community of practice approach. *Technical Communication, 59*(1), 32–48.

44 Fisher, L., & Bennion, L. (2005). Organizational implications of the future development of technical communication: Fostering communities of practice in the workplace. *Technical communication, 52*(3), 277–288.

45 Spilka, R. (2010) *Digital literacy for technical communication: 21st century theory and* practice. Routledge.

46 Pirsig, R. M. (1974). *Zen and the art of motorcycle maintenance: An Inquiry into Values.* Bantam Books.

47 Savage, G. J., & Sullivan, D. L. *Writing a professional life: Stories of technical communicators on and off the job.* Allyn & Bacon.

48 Bridgeford, T., Kitalong, K. S., & Williamson, B. (2016). *Sharing our intellectual traces: Narrative reflections from administrators of professional, technical, and scientific programs.* Routledge.

49 Wizeline. (2018, March 12). *A day in the life of a technical writer* [Video]. YouTube. https://www.youtube.com/watch?v=Tye1AnSopo4

50 Institute of Scientific and Technical Communicators. (n.d.). *A day in the life of a technical communicator.* https://istc.org.uk/homepage/value-of-technical-communication/a-day-in-the-life-of-a-technical-communicator/

51 Johnson, T. (2016, September 2). Balancing the never-ending list of documentation to write with your natural interests and passions. *I'd Rather be Writing*. https://idratherbewriting.com/2016/09/02/the-never-ending-list-of-tasks-to-complete/

52 Technical Writing is Easy. (2019, March 3). *A typical day for a technical writer*. https://medium.com/technical-writing-is-easy/a-typical-day-for-a-technical-writer-5c82cabb5e33

53 Hart-Davidson, W. (2013). What are the work patterns of technical communication? In J. Johnson Eilola & S. A. Selber (Eds.), *Solving problems in technical communication* (pp. 50–74). Chicago University Press.

54 Brady, A. (2003). Interrupting gender as usual: Mêtis goes to work. *Women's Studies, 32*(2), 211–233. https://doi.org/10.1080/00497870310061

55 Edenfield, A. C. (2018). The burden of ambiguity: Writing at a cooperative. *Technical Communication, 65*(1), 31–45.

56 Colton, J. S., & Holmes, S. (2018). A social justice theory of active equality for technical communication. *Journal of Technical Writing and Communication, 48*(1), 4–30. https://doi.org/10.1177/0047281616647803

57 Petersen, E. J. (2017). Articulating value amid persistent misconceptions about technical and professional communication in the workplace. *Technical Communication, 64*(3), 210–222.

58 Brumberger, E., & Lauer, C. (2020). A day in the life: Personas of professional communicators at work. *Journal of Technical Writing and Communication, 50*(3), 308–335. https://doi.org/10.1177/0047281619868723

59 Henry, J. (2000). *Writing workplace cultures: An archaeology of professional writing*. Carbondale, IL: Southern Illinois University Press.

60 Wilson, G. (2001). Technical communication and late capitalism: Considering a postmodern technical communication pedagogy. *Journal of Business and Technical Communication, 15*(1), 72–99. https://doi.org/10.1177/105065190101500104

61 Wilson, G., & Wolford, R. (2017). The technical communicator as (post-postmodern) discourse worker. *Journal of Business and Technical Communication, 31*(1), 3–29.

62 Lanier, C. (2018). Toward understanding important workplace issues for technical communicators. *Technical Communication, 65*(1), 66–84.

63 Carliner, S., & and Chen, Y. (2018, December). Who technical communicators are: A summary of demographics, backgrounds, and employment. *Intercom, 65*(8), 8–12.

64 Jones, N. N., Moore, K. R., & Walton, R. (2016). Disrupting the past to disrupt the future: An antenarrative of technical communication. *Technical Communication Quarterly, 25*(4), 211–229. https://doi.org/10.1080/10572252.2016.1224655

65 Matheson, B., & Petersen, E. J. (2020). The profession of technical communication through the lens of the STC India Chapter: Understanding current perspectives and future directions. *Technical Communication, 67*(3), 25–43.

66 Virtaluoto, J., Sannino, A., & Engeström, Y. (2016). Surviving outsourcing and offshoring: Technical communication professionals in search of a future. *Journal of Business and Technical Communication, 30*(4), 495–532. https://doi.org/10.1177/1050651916651908

67 Brumberger, E., & Lauer, C. (2015). The evolution of technical communication: An analysis of industry job postings. *Technical Communication, 62*(4), 224–243.

68 Lauer, C., & Brumberger, E. (2016). Technical communication as user experience in a broadening industry landscape. *Technical Communication, 63*(3), 248–264.

69 Brumberger, E., & Lauer, C. (2017). International faces of technical communication: An analysis of job postings in three markets. *Technical Communication, 64*(4), 310–327.

70 Watts, J. (2020). Fostering industry connections through workplace-situated graduate student research. *Technical Communication, 67*(3), 80–102.

71 Stanton, R. (2017). Do technical/professional writing (TPW) programs offer what students need for their start in the workplace? A comparison of requirements in program curricula and job ads in industry. *Technical Communication, 64*(3), 223–236.

72 Melonçon, L., & Henschel, S. (2013). Current state of US undergraduate degree programs in technical and professional communication. *Technical Communication, 60*(1), 45–64.

73 Boettger, R. K., & Friess, E. (2016). Academics are from Mars, practitioners are from Venus: Analyzing content alignment within technical communication forums. *Technical Communication, 63*(4), 314–327.

74 Boettger, R. K., & Friess, E. (2020). Content and authorship patterns in technical communication journals (1996–2017): A quantitative content analysis. *Technical Communication, 67*(3), 4–24.

75 St.Amant, K., & Melonçon, L. (2016). Reflections on research: Examining practitioner perspectives on the state of research in technical communication. *Technical Communication, 63*(4), 346–364.

76 Rude, C. D. (2015). Building identity and community through research. *Journal of Technical Writing and Communication, 45*(4), 366–380. https://doi.org/10.1177/0047281615585753

77 Petersson, P., Springett, J., & Blomqvist, K. (2009). Telling stories from everyday practice, an opportunity to see a bigger picture: a participatory action research project about developing discharge planning. *Health and Social Care in the Community, 17*(6), 548–556. https://doi.org/10.1111/j.1365-2524.2009.00854.x

8   Data Gathering and Analysis

Introduction

Any research study is complicated and has many considerations. Because this project had a broad focus on both the profession and practice of technical communication, it had the potential to be especially complicated. To help me to recognize and integrate variables, I conducted a situational analysis. This step helped me to identify aspects of technical communication that I might otherwise have overlooked.

The primary data source I used consisted of narratives from practitioners, referred to throughout the book as practice narratives. Individual practitioners described their educational backgrounds, work experiences, and future expectations for technical communication. Because this data source provided only snapshots of the profession and practice of technical communication, the book's content is based on two additional, complementary sources:

This chapter begins with a discussion of my **motivations** in undertaking the study described in this book.

Whenever we conduct research, we make decisions, some of which are obvious to readers, others, perhaps less so. This chapter explains the **decisions I made when gathering and analyzing data** for this book.

It sets out the research questions and explains how I gathered and analyzed data to help me to respond to them. It outlines the **three phases** of the study and also describes some limitations that future studies could address.

- A literature review of secondary research.
- An analysis of online resources.

This chapter will help you to recognize the research decisions I made and to trust their outcomes. You may find the description of the study helpful for designing and explaining your own research projects.

- If you are a **technical communication student**, this chapter will guide you through the steps I took to design the study. The chapter will help you to identify and analyze stages in a research process. It will serve as a guide for setting up a similar project.
- If you **work in industry as a technical communicator**, this chapter will enable you to match the methods and analytical strategies with the contents of previous chapters. The chapter will help you to trust the book's content.
- **If you are a teacher or a researcher**, you can use the contents of this chapter as a case study that explains the design of a qualitative research project.

Background: Why I Conducted this Study

Part of doing a narrative study is acknowledging that you are part of the story. As researchers, we are "part of the storied landscapes we are studying" (p. 25).[1] Connelly and Clandinin explain that "the stories of who we are as researchers are also evident as we compose our research texts" (p. 20).[2] In that spirit, I would like to explain how I came to do this project and how my background influenced the book.

I have taught technical communication at the University of Limerick, in Ireland, since 1999. After completing a master's degree in technical communication, I was hired as a lecturer (junior professor) because no applicants for the post had a doctorate in a relevant area. (Over 20 years on, these posts are much more competitive.) I began my doctoral work in 2003. The challenge of simultaneously teaching and doing a big research project influenced both the approach to and the subject of my research. Although I had some industry experience, I was eager to learn as much as I could about technical communication from industry practitioners. I was aware, from my interactions with alumni and employers, of the need for a better understanding of the profession and practice of technical communication in Ireland. I was simultaneously influenced by the status of technical communication in my university at the time. I was a junior member of staff in a small subject within a big department. The main areas of concern in the department were modern languages and literature, and technical communication was pretty peripheral. Finding myself somewhat alienated from my colleagues' teaching and research activities led me to become interested in whether, and in what ways, the low profile of technical communication reported in industry affected practitioners.

I chose Ireland as the focus of my doctoral study for convenience but also because various geographic, economic, and social factors influenced the development of the profession and practice of technical communication in this country. At the time, Ireland was one of only two European Union countries where the first language was English. We had a very limited history of manufacturing; Ireland was a primarily agrarian economy until the 1970s. In the 1990s, many information and communications technology and pharmaceutical companies located offices in Ireland because of low corporation taxes, inward investment, relatively low wages, and a workforce that was young, educated, and English-speaking. Because of these factors, the Irish technical communication sector was based primarily in the software industry. Technical writers worked in large teams or as lone writers in multinational corporations. A small number of technical communication service companies also employed technical writers. Because the profession was new, we did not have a professional association or a local chapter of an international association. A couple of attempts to establish a professional association had failed.

In the mid-2000s, professionalization was an active topic in technical communication research. As you saw in Chapter 7, many researchers were exploring how technical communication could increase its status and become more professionalized. The concept of communities of practice was also influencing many types of workplace research.

My Doctoral Study

My experiences of teaching technical communication in Ireland led me to design a study that explored Irish technical communicators' perceptions of their profession and practice and that had theoretical bases in the intersections of professionalization theory and communities of practice theory. The intersections I examined were education and training, status and value, practice, and community. I gathered data from practitioners in Ireland through surveys, interviews, and an online focus group. After I competed my doctorate, I took the findings from that study into the classroom. Students were energized and empowered by what they learned about practitioners' work, experiences, challenges, and communities. Sharing these anonymized accounts of practice with students was much more effective than trying to explain abstract theories.

Taking a Global Perspective

Although my doctoral findings helped students to understand practice contexts, concerns, and concepts, the findings were becoming dated, and they were specific to working in Ireland. I became convinced of the need for a book about the profession and practice which would be relevant and valuable for other students, other teachers, and indeed practitioners. Technical communication practice is heterogenous, and many practitioners would benefit from knowing more about activities in workplaces other than their own.

A useful book would need to be relevant not only in Ireland but globally because technical communication is a global and globalized profession. Many of the concerns of practitioners globally are similar, but there are also local differences and nuances. To broaden the scope of my doctoral study, I gathered data from technical communicators in 13 countries, and I also used international websites, reports, academic articles, and social media to ensure that the book was as current, expansive, and inclusive as possible.

The remaining sections of this chapter describe how I conducted the study and the phases of this project. I have outlined them here to enable you to understand my process and to adopt a similar process, or adapt it, if you are conducting a similar qualitative study.

Phase 1: Planning the Study

Research projects must be carefully planned and systematic. A researcher needs to identify the approach, methods, and research questions.

The Research Approach

This is primarily a qualitative project. This study focuses on people's experiences. My expertise is in qualitative methods like interviewing and focus groups, and the methods I used in this project collected words rather than numbers. Although a couple of chapters include descriptive statistics that emerged from the narrative data, these are intended only to suggest patterns that a future quantitative or qualitative study could explore in significantly more depth. The number of participants in the study was too small for me to draw any meaningful statistical inferences. Moreover, qualitative studies work well in highlighting the emic perspective: the insider experiences that were so important to this book.

I applied **symbolic interactionist principles** (discussed in Chapter 7) to gather and analyze data from technical communicators. The study described in this book sees technical communicators as workers responding to fluctuating work organizations, practices, and technologies. Their perceptions of their occupation depend on:

• Their interactions with one another and with other occupations.
• The processes of community formation.
• The meanings they attribute to their practice and their roles.

This project incorporates narrative inquiry. Narrative inquiry is "a way of honoring lived experience as a source of important knowledge and understanding" (p. 17).[1] I chose to gather practitioner narratives because I was aware, from both research and teaching experiences, of how convincing and compelling stories can be. For example, when practitioners present guest lectures, students are engaged by their authentic insights into their work: insights that suggest an overarching career trajectory, that dramatize, that personalize, and that enable students to empathize and to visualize their own futures. The narratives I gathered for this study also enabled practitioners to describe their experiences in their own words and ensured they had a prominent voice within the book.

I analyzed the data using a **modified grounded theory** approach. Grounded theory is widely used in qualitative research because it enables researchers to generalize common patterns from research data that are primarily words.[3] In grounded theory projects, researchers generate theory from patterns they find in data. Because I had completed a doctorate in a related area, I was already familiar with theory and with other studies in this field. For this reason, this study was only partially grounded in data. My prior knowledge of theory made it a modified grounded theory study. I used tools of grounded theory such as coding, keeping memos, and discerning patterns from words. I discuss these aspects in more detail in the "Analyzing and Reporting the Practice Narratives" section. Situational analysis is a grounded theory method[4] that was an important part of my process.

Situational Analysis

Situational analysis can involve three mapping approaches: situational maps, social worlds/arenas maps, and positional maps. For this study, I developed only a situational map.

Before I identified research questions or began to gather data, I conducted a situational analysis. I was influenced by the work of Adele Clarke, in particular her book *Situational Analysis: Grounded Theory after the Postmodern Turn*.[4] Clarke classified situational analysis as a first phase in a grounded theory project, where researchers analyze the research context (situation) and map it as explicitly as possible. This step helps researchers to ensure that they are being comprehensive and systematic. Situational analysis is suitable for dynamic settings, such as occupational studies. A researcher maps everything they can think of that is relevant to the research project and attempts to determine how the various elements are related. Situational analysis complements a symbolic interactionist perspective in research because the researcher has to consider how the elements in the map interact with one another.

Purpose of a Situational Map

A situational map should help a researcher to decide what is important for the study and how to approach the research problem systematically. You can use situational maps to determine which data to collect and how to sort "the vast amounts of data that one 'uploads' into one's brain and other sites during the research process" (p. 85).[4] Maps can include human elements (like individuals, groups, organizations, institutions, and subcultures) and nonhuman elements (like infrastructure, artifacts, and technology). Developing the map is a kind of brainstorming activity, and you can use mind maps or similar tools to structure the map. There is no one "correct" map. "What appears in your situational map is based on your situation of inquiry – your project. The goal here is not to fill in the blanks but to really examine your situation of inquiry thoroughly" (p. 89).[4]

Situational analysis broadens the site of study to the entire situation. Taking a broad perspective helped me to consider "heterogenous situations and relations" (p. xxiv)[4] rather than trying to simplify data. Because this book explores both the profession and practice of an entire occupation, it made sense to take a broad perspective. Broad mapping is also appropriate for situations that might be considered postmodern. As we saw in Chapter 7, technical communication has been described as postmodern because it is a dynamic field of work.

Situational Map: The Profession and Practice of Technical Communication

To begin creating my map, I followed guidance from Clarke (p. 87)[4] and asked:

- Who and what are in this situation?
- Who and what matter in this situation?
- What elements "make a difference" in this situation?

My situational map is depicted in Figure 8.1. As you can see, I aimed to include as many facets of the situation as I was aware of and to catalog the variables of the project. This map of the technical communication profession and its practice formed the basis of data gathering and analysis for this book.

This map includes the following:

- Individuals whose contributions I believed would be relevant: practitioners and managers who responded to the narrative survey and academics, leaders and influencers whose publicly available work (blogs, trade publications, social media, and academic publications) is relevant.
- Communities: project groups, social media groups, committees, and professional associations.
- Public discourse: online content, trade publications, and academic publications that would contribute to my understanding of the situation and that would help me to analyze the data.
- Practices, organizations, hot topics, and sociocultural/intercultural/political influences: These lists could expand, contract, or change in various ways during the study.
- One part of the map illustrates my own background and how this might influence the project. This part was meant to remind me of my preconceptions and to help me to avoid any biases that these preconceptions might introduce.

The map represents the research site as I saw it at the beginning of the project and that guided my approach to data gathering. It was based on my initial understanding, before conducting the study, of the issues that I believed would impact on the technical communication profession and practice and that would affect this study. In retrospect, it was, of course, incomplete.

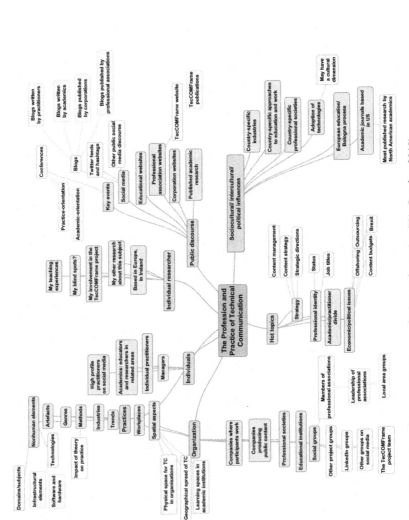

Figure 8.1 Situational analysis of the technical communication profession and practice (V1: October 2019)

- I did not include some items that later proved to be important. For example, diversification and a multiplicity of job titles constitutes an important professional and strategic theme that did not feature in the map.
- Some items that I included in the map, meanwhile, proved to be outside the scope of the study or did not feature in the data. For example, few participants mentioned political or geographical influences. Owing to the limitations of the sample size and limited research on this topic, the book does not explore country-specific sectors or situations. In addition, blogs written by academics did not feature in my content analyses.
- Lastly, it was impossible for me to predict some situations that would become relevant. A situation that emerged when I was analyzing the data and starting to write up the study was the Covid-19 pandemic and its impact on almost every aspect of life throughout the world, including technical communication and how it is practiced.

Nevertheless, the map was an important starting point. Mapping the situation as I saw it at the beginning of the project helped me to consider the variables that could be relevant to technical communicators:

- The global context in which technical communicators operate.
- Educational opportunities.
- Individual workers.
- Professional and social communities.
- Related professions and professionals.
- Technology.

It helped me to define the scope of the book. Later, it became a touchstone that I came back to during the project. Even when my focus changed and some elements became irrelevant while others emerged, the map helped me to track my progress and forced me to reflect on changes in direction.

> If you are embarking on a big research project, with a broad scope that you need to pin down, situational analysis may be a very helpful tool.

The Research Questions

My objectives in this study were to examine the profession and practice of technical communication globally, use a qualitative and narrative approach, and base the study on a broad map of potential influences. These were the research questions upon which I based the data gathering. In parentheses, you can see the parts of the book where I have responded to these questions:

1. How do technical communication practitioners throughout the world characterize their profession, education, communities, work, and workplaces? (I respond to this question in Chapters 1–5.)

2. How does contemporary English-language public, trade, and academic discourse about technical communication represent the profession and its practice? (I respond to this question in Chapters 1–5.)
3. What do these and other sources suggest for the future of technical communication? (I respond to this question in Chapter 6.)
4. What strategic and research directions do these sources implicitly or explicitly identify for the profession and practice of technical communication? (I respond to this question in Chapter 9.)

These questions map to Carolyn Rude's[5] overarching research question for our profession (presented in Chapter 1): *How do texts (print, digital, multimedia; visual, verbal) and related communication practices mediate knowledge, values, and action in a variety of social and professional contexts?* My research questions also connect with three research strands that she identified: disciplinarity, pedagogy, and practice.

Phase 2: Gathering the Data

To respond to the questions, I undertook these activities:

- Conducted a literature review.
- Designed and distributed a survey to gather practice narrative data.
- Conducted content analyses of additional (primarily online) sources.

Literature Review

All researchers stand on the shoulders of giants. We base our studies on theories developed by other researchers, take up their invitations to follow a research direction, and build upon their findings. This book is no different. Throughout, I have interpreted and built upon other researchers' work.

- Each chapter incorporates analysis of published research relevant to that chapter's focus.
- Chapter 7 presents the theoretical foundation of the book, explaining how I adapted theories of professions and practice to structure and interpret the book's contents. This chapter also discusses how I interpreted and built upon relevant research about industry perspectives on the profession and practice of technical communication.

The literature review process for this book was iterative. Before I began the project, I was already reasonably familiar with some of the published research, from teaching and research. But the situational map was broader than any previous studies I had undertaken, and the theoretical framework (intersections of professions and practice) had also evolved since I completed my doctorate. Furthermore, new academic and trade articles are published all the time. I have done my best to be expansive and comprehensive in my literature search. As any researcher will attest, however, a literature review is never finished. No doubt, I have missed important articles, books, or other sources.

Judging from the literature review, I ascertained that the current profession and practice-focused research in technical communication has some limitations to which I hope this book begins to respond (these limitations are discussed in more detail in Chapter 7):

- We have a dearth of industry-based research into the profession in the past decade.
- There is a lack of correspondence between what practitioners need from research and the research that academics do.
- Many studies have a North American focus.
- There is a lack of conversation about how research, especially research with practitioners, can contribute to the strategic development of technical communication.

Practice Narratives

To begin to address these limitations, I gathered narrative data from industry practitioners about their educational experiences, previous and current work in technical communication, and their future expectations for their profession. Narrative inquiry is "a way of understanding experience. It is a collaboration between researcher and participants over time, in a place or series of places, and in a social interaction with milieus. [...] narrative inquiry is stories lived and told" (p. 20).[2] I chose a narrative approach to enable industry practitioners to construct their stories and to represent themselves.[6] Narratives have immense power to reveal the nuances and hidden characteristics of a situation. Narratives are, of course, also subjective and describe individual interpretations of events and situations. For these reasons, narrative research "stands in important contrast to traditional research in which investigators test hypotheses about abstract principles" (p. 66).[6] Although some narrative studies have been conducted in technical communication (see Chapter 7), I believe that the experiences, perspectives, and voices of industry practitioners are under-represented in our academic work. Bakhtin[7] encourages us to listen to all voices, not just those of the powerful or prominent.

How I Gathered the Narrative Data

I gathered narrative, qualitative, reflexive, and reflective personal accounts of work experiences from practitioners through a qualitative survey that was approved by my university's Research Ethics Committee (REC) (our equivalent of the Institutional Review Board). I designed the survey to give practitioners space and scope to provide anonymous and confidential input about their education and work histories, current workplaces and activities, and future projections for the profession. The questions had both "a critical and creative function" (p. 68)[6] because the input from practitioners had the scope to challenge my assumptions about the nature of the profession and its practice.

You can access the survey and the documentation I submitted to the REC in Appendix B.

Survey Design and Distribution

The survey had seven questions: three that sought basic demographic information and four that were open-ended and requested narrative responses. The three demographic questions were about gender, age range, and location. The purpose of these questions was to enable me to determine the range and variety of responses. The four qualitative questions asked respondents for narrative responses about their:

- Educational background.
- Career history.
- Current role.
- Vision and preparations for the future of technical communication.

I asked respondents to use their own words to describe what they believed to be important factors in their experiences. The questions included prompts to focus the responses, but respondents were not required to include the prompt categories in their responses. No questions compromised anonymity. For example, I did not ask respondents their names, the names of the organizations where they worked, or any other personal details. I took various measures to ensure that the survey was secure and confidential and that I handled data in compliance with the General Data Protection Regulation. (You can find more about these measures in Appendix B.)

I used a mix of convenience and snowball sampling to gather survey responses. I shared the link on several technical communication LinkedIn groups. I also shared it with professional associations, special interest groups, and my personal contacts. Whenever I shared it, I asked the recipients to share it further within their networks.

The purpose of the narratives was not to build a definitive set of profiles but rather to gather snapshots of possible educational routes and career paths, to learn more about day-to-day practice, and to get an impression of what practitioners saw for the future for the profession.

Why I used a Survey to Gather the Narrative Data

I chose a survey instrument for several reasons:

- From a practical perspective, it was an inexpensive and convenient way to gather large amounts of data from as many participants as possible.
- It was an acceptable data-gathering instrument for my university's REC.
- It had the potential to reach a global pool of respondents.
- It enabled anonymous and confidential participation.

- Respondents could take as long as they needed to respond.
- People are familiar with this type of instrument. We complete surveys all the time in our work and personal lives. I hoped that this familiarity would encourage respondents to participate.
- The open-ended narrative questions reduced potential researcher bias.[8] Although I included some prompts within each narrative question, participants were free to respond in ways that were relevant or meaningful to them.

Of course, surveys also have many limitations. In the case of this survey, I identified these limitations:

- Owing to the request to provide narrative data for four questions, the survey took a long time to complete. This factor almost certainly reduced participation. Pilot participants spent an average of 40 minutes answering all questions. They were all native English speakers. I had not considered how much more time-consuming the survey would be for non-native speakers writing their responses in English. The final analytics showed that participants had spent an average of 147 minutes completing the survey. This was a huge demand on busy people and likely impacted participation.
- I was not able to follow up with participants about statements I did not understand or wanted to know more about.
- The data were time-consuming to analyze, a feature of any large qualitative study.

Content Analyses

The participants who responded to the survey chose how to respond and which aspects of their work to emphasize. Because the narratives were snapshots of individual experiences, they were not generalizable. In order to expand the account of the profession and practice, I looked for gaps in the data and sought to fill these gaps by analyzing several contemporary, mostly online, primarily English-language, sources, among them sources authored by industry practitioners. These sources included the following:

1) Online content about education and practice published by technical communication associations and practitioners, *inter alia,* public social media feeds, podcasts, webinars, blog posts, discussion forums, newsletters, and corporate websites.
2) Trade magazines and online content from and about technical communication professional associations, including the following:
   - Society for Technical Communication.
   - Institute of Scientific and Technical Communicators.
   - tekom Europe.
3) Online content from academic providers of technical communication programs in Europe and North America.

The content analyses were meant to:

- Supplement the literature review with content that had a stronger industry focus.
- Extend the global reach of the practice narrative data.

The online content sources provided up-to-date information about a variety of professional and practice concerns, including education, activities, workplaces, and future trends.

Phase 3: Analyzing and Reporting the Practice Narratives

The practice narrative survey was open in October and November 2019. The survey tool I used (Microsoft Forms) reported basic demographic data. There were 62 responses from individuals in 13 countries: 26 identified as male and 36 identified as female. Although individual respondents did not have to answer every qualitative question, most respondents did, as follows:

- 57 wrote a narrative about their education.
- 58 wrote a narrative about their work history.
- 59 wrote a narrative about their day-to-day work.
- 57 wrote a narrative about their expectations and preparations for the future.

Although the analytic approach was primarily qualitative, I was able to generate some descriptive statistics, which are reported occasionally (e.g., the percentage of respondents with graduate degrees). Where it was useful to do so, I cross-tabulated demographic and narrative survey responses to determine potential patterns (e.g., education levels of respondents within an age range). I have reported these patterns only to indicate where they might suggest directions for a focused study. Since the number of respondents was small and the data were primarily qualitative, cross-tabulation was generally not an appropriate analytical strategy.

Qualitative Analysis: Categorizing and Coding the Practice Narratives

I analyzed the narrative data using qualitative approaches adapted from grounded theory. These were the steps I undertook, following Hughes and Hayhoe's analytic steps for qualitative research.[8] The process was iterative and I cycled back and forth between steps throughout the analysis:

1. I copied transcripts of the responses to each narrative question into individual Microsoft Word documents.
2. I read each document twice, highlighting key words on the second reading.
3. Next, I coded the data. I had some predefined codes that were linked to the survey questions (e.g., education). Others were grounded in the data (e.g., feelings about work). I refined the codes as I proceeded.

4. During the coding process, I kept memos of my thoughts, using Microsoft Word comments. This step helped me to reflect, ground my interpretations in the data, and capture conflicts and insights as they occurred to me.[3]

5. The next step was to identify prominent categories where I saw patterns emerging from the coding. I used Excel to catalogue excerpts that corresponded to each category. I used the constant comparative method that Glaser and Strauss[3] recommend: comparing every excerpt in a category and across categories to be sure I was being consistent.

To protect the anonymity of individual participants, I have reported excerpts from the narratives rather than complete contributions. If respondents provided data that could identify them (e.g., an organization's name), I deleted those data and I did not include them in the book.

Writing the Vignettes

A somewhat unusual feature of this study is the vignettes that are included in several chapters. Although some technical communication websites and recent articles use personas or profiles of technical communicators, vignettes provide more detail about a singular situation or scenario. I developed the vignettes following the definition of Ely et al. (p. 70).[9] Vignettes are:

> compact sketches that can be used to introduce characters, foreshadow events and analysis to come, highlight particular findings, or summarise a particular theme or issue in analysis and interpretation. Vignettes are composites that encapsulate what the researcher finds through the fieldwork.

Vignettes are typically used in two ways in research studies:

1. Some researchers use them to gather data based on the participants' responses to reading a vignette. This way of using vignettes is reasonably common in medical studies, for example.

2. They can also be used to represent key findings of a study. Wenger used them in this way in *Communities of Practice*[10] to illustrate his ethnographic research findings about how practice became integral to people's lives.

I have taken the second approach. The vignettes in this book are short stories that illustrate a situation, trajectory, or environment. They were composed on the basis of the practice narrative responses: every specific piece of information in each vignette came from the practice narratives, but each individual vignette is a composite. The purpose of the composition process was to increase the representativeness of the vignettes while protecting the anonymity of individual respondents. I also sought to ensure that the vignettes are culturally neutral by removing culture-specific references or terminology. I used respondents' own words, as much as possible. I rewrote content occasionally to improve the narrative flow and when respondents' words could identify them personally (e.g., a couple mentioned specific education programs and organizations).

The purpose of the vignettes was to:

- Bring education and practice to life for readers.
- Make the data relatable to practitioners.
- Help newcomers and lateral entrants to technical communication visualize their future involvement in the field.
- Represent as many practitioner voices and experiences as possible in the book.

Because they were composed from practitioner contributions, they represent an authentic record of various aspects of the profession and practice. Although they are incomplete and unrepresentative at a general level, the vignettes are "one account of the truth that was representative of events" (p. 954).[11] To an extent, they are staged versions of reality, and I connect them back to Goffman's work on performance and dramatization. Ely et al. acknowledge that, although this staging of events can cause difficulty for researchers, who may fear misrepresenting the participants through generalizing their stories, vignettes can "reveal implicitly the significance of the story told" (p.72).[9] In this book, for example, they suggest ways in which individuals entered technical communication, engaged with the field and its practice, and are currently working. Because they are composites, they incorporate my analysis of the narratives.

Limitations of this Project

Like any project, this study has limitations. I believe that the book's wide lens was necessary because of a lack of similar texts and a dearth of recent research about the profession as a whole. Nevertheless, the broad scope limited my ability to follow several interesting threads. There are other ways of looking at all of these topics, and because of the wide lens, I skimmed over some themes rather than discussing them in depth.

Another consideration, which I do not see as a limitation, is that the record of technical communication presented here is of its time. Social constructionists and narrative theorists agree on the impact of temporality on qualitative research. I believe that it is essential to record the variety and distinctive character of technical communication and that now is an important time to do so. In a shifting landscape, students, teachers, and practitioners need more research-based information about practice and about the profession. Furthermore, in spite of the potential for this content to date, several themes have persisted for decades. They include diversification and continual changes wrought by technology.

Initially, I hoped to be able to write detailed accounts of technical communication practice in different countries. As I explored the situation, it became clear to me that because I planned to gather qualitative rather than quantitative data, this goal was too ambitious. I would not be able to develop profiles of technical communication in different countries and within various sectors in those countries. Rather than writing *about* international technical communication contexts, I have included voices from practitioners internationally. Even so, this project did not include Asian, African, or South American participants. These are limitations that I hope future studies will address.

Summary of Chapter 8

This study builds on my doctoral work that explored technical communication practice in Ireland. I built on that study to update and internationalize the findings.

- The situational map represents the initial scope and perspective of the book and is based on my interpretations and analysis of the most important human and nonhuman characteristics of the profession and practice.
- I used three data sources: academic articles, practice narratives, and content analyses of online sources.
- This book is based on a primarily qualitative study, and I took a modified grounded theory approach to data analysis.
- An unusual feature of the book is the vignettes. I composed these from the practice narratives. They were conceived to represent generalized but authentic records of technical communication practice.
- This study has some limitations to which future studies can respond.

Discussion Questions

1. The situational map shows the various elements that I thought mattered or might matter in this project. Develop a situational map for a project you are currently working on. Aim to make it as detailed as possible. Consider who and what are in this situation. In what ways do they make a difference to the situation?
2. The vignettes are an unusual feature of this study. Search for a recent article that uses vignettes to represent research findings. What discipline is the article from? How were the vignettes written and reported?
3. What other methods could I have used to gather data for this project? What methods would you use? How would you undertake a similar study?
4. How does the description of the study outlined here help you to understand the research process? Do any features of the process surprise you?

References

1 Clandinin, D. J., Cain, V., Lessard, S., & Huber, J. (2016). *Engaging in narrative inquiries with children and youth.* Routledge.
2 Connelly, F. M. and Clandinin, D. J. (2000). *Narrative Inquiry: Experience and story in qualitative research.* Jossey-Bass.

3 Glaser, B. G., & Strauss, A. L. (1999). *The discovery of grounded theory: Strategies for qualitative research.* Aldine de Gruyter.

4 Clarke, A. (2005). *Situational analysis: Grounded theory after the postmodern turn.* Sage.

5 Rude, C. D. (2009). Mapping the research questions in technical communication. *Journal of Business and Technical Communication, 23*(2), 174–215. https://doi.org/10.1177/1050651908329562

6 Gergen, K. (2009). *An invitation to social construction.* Sage.

7 Bakhtin, M. M. (1981). *The dialogic imagination: Four essays.* (C. Emerson, Trans., & M. Holquist, Ed. & Trans.). University of Texas Press. (Original work published 1975)

8 Hughes, M. A., & Hayhoe, G. F. (2008). *A research primer for technical communication: Methods, exemplars, and analyses.* Lawrence Erlbaum Associates.

9 Ely, M., Vinz, R., Downing, M., & Amzul, M. (1997). *On writing qualitative research: Living by words.* Routledge.

10 Wenger, E. (1998). *Communities of practice: Learning, meaning, and identity.* Cambridge University Press.

11 Spalding, N. J., & Phillips, T. (2007). Exploring the use of vignettes: From validity to trustworthiness. *Qualitative Health Research, 17*(7), 954–962. https://doi.org/10.1177/1049732307306187

9 Conclusions and Future Directions

Introduction

The profession of technical communication has a distinctive character but that distinction may be less obvious because aspects of the practice change continually and often this is due to external factors such as inevitable shifts in economic patterns and technology. Given the complexity and diversity of this profession and practice,

In this chapter, I attempt to draw out the **key insights** from the findings, and I examine the potential for these findings to shape **future research** and **strategic directions** in technical communication.

neat conclusions are elusive. Instead, I propose to tell you my interpretations of the main findings of this book and what I believe the findings mean for research and strategic planning in technical communication.

Insights about the Profession and Practice of Technical Communication

As we have seen throughout the book, technical communication is a developing and dynamic occupational field. Over the past two decades, however, it, like many professions, has become fragmented and has diversified and this is partly due to globalization and digitalization. Diversification and fragmentation are evident, for example, in the variety of job titles that come under the technical communication umbrella. Our work patterns, too, exemplify variety of competencies, activities, artifacts, technologies, and genres.

Our educational trajectories are also characterized by diversity. Technical communicators come from many educational backgrounds. This is still likely to be a second or later career for many of us, and our primary degrees are rarely in technical communication. In many countries, no educational programs exist, and academically technical communication is most advanced in the US. Perhaps because of our diverse educational backgrounds, professional development is important to practitioners, and many professional development opportunities exist. Academic and training projects in recent years are also helping to shape our educational and professional development landscape.

Multiple communities support technical communicators, from professional associations, to online communities, to communities of practice that are sometimes workplace-specific. Communities and professional associations support our work in practical ways. Although some professional associations support both academics and industry practitioners, the interactions between these groups are limited at present. Furthermore, the profusion of communities may point toward an ambiguous professional identity and a lack of agreement about where we belong.

Although technical communicators can work in any industry sector as well as in public and nonprofit organizations, we are most likely to work in information and communications technology and manufacturing environments in private industry. Different work profiles and work environments suit individuals at different career stages, and we have opportunities to work from home. This profession has a very positive job outlook internationally. Various data sources, including the practice narratives, indicate a high level of job satisfaction and potential for job mobility. Because we tend to work in corporations, however, we may be vulnerable to global corporate practices like restructuring, offshoring, and outsourcing.

Many work patterns, including emerging skills, new labor models, and technology transformations, already affect technical communicators. Although several of our competencies will endure, especially our problem-solving, adaptability, and writing and rhetorical skills, we need to develop new competencies. These competencies are dictated, at least partially, by work sectors. The future of technical communication, like its past, seems likely to be heavily influenced by technological advancements.

The value of clear communication and the value of our work in planning, creating, and enabling access to information are more important than ever before. The global lockdowns and stay-at-home orders of 2020 underscored the importance of our practice, as people turned to online platforms to work, collaborate, learn, and access information.

Persistent Concerns: Strategic and Research Directions

As presented in Chapter 8, my final research question for this study was:

> What strategic and research directions do these sources [the practice narratives, literature review, and content analyses] implicitly or explicitly identify for the profession and practice of technical communication?

In earlier chapters, I noted some specific and recognized research and strategic challenges for technical communication:

- As the profession has become more fragmented, our professional identity has become ambiguous. Research into the profession has diminished in the past decade, coinciding with a move away from professionalization research in sociology. We have not found an alternative way to conceptualize research about our profession, our professionalism, or our strategic direction, however. We lack a coherent research thread that focuses on the development of our profession and its practices.[1] Perhaps for this reason, we also lack a clear developmental agenda. This research thread and agenda were prominent in the past.
- This profession emerged and developed more quickly in North America than in other regions of the globe. This has resulted, perhaps inevitably, in a situation where most research, indeed most published information, about this profession comes from North American sources. Not only that, Boettger and Friess concluded from a study of journal articles published between 1996 and 2017 that "the majority of the field's research is conducted at only a handful of similar academic institutions concentrated in the US" (p. 21).[2]
- Industry practitioners have online and face-to-face networks separate from those of academics. This separation has been ongoing for some time and has led to a situation where many of our research studies do not seem to correspond with what practitioners need from research. Research projects by practitioners and by academics "exist in different spheres, as though two conversations are taking place that have little to do with each other" (p. 373).[3] This separation may limit curricular, strategic, and research development.

The sense that technical communication is under threat or needs a strategic direction is not new. Giammona[4] discussed the future of technical communication in a 2004 article that analyzed primary research with thought leaders. She concluded that technical communicators needed to think "outside the box" to thrive. Covid-19 has shown us the importance of our work and may have presented us with the inflection point we need to find and adopt such creative solutions.

What are these creative solutions? Boettger and Friess[5] concluded a 2016 article about the disconnect between academic research and industry practice with three calls to action:

1) To "unify existing forums" (p. 322).
2) To "identify new audiences" (p. 323).
3) To "engage practitioners in research" (pp. 323–324).

Building on those calls to action and on the earlier chapters of this book, I attempt to respond to the final research question by suggesting ways we might proceed. I recognize, however, that these challenges have been with us for some time and that there are no obvious or easy solutions.

Strengthening our Professional Identity

Among the most important strategic considerations in technical communication are whether and how to examine and address our ambiguous professional identity. Technical communicators' professional consciousness "seems to have eroded over time" (p. 322),[6] partly due to the many technical communication roles, communities, and program and job titles. Labels are changing reflexively at local levels in industry and academia. As we have become more diverse and more titles have come into use, we have become less recognizable, even to one another. St.Amant and Melonçon argued that technical communication is "doom[ed] ... to fail unless we can change the field's perspective of what we consider *common ground*" (p. 4).[1]

Systems and institutions (e.g., professional associations) have a big impact on our profession and practice.[7] A strong professional identity can increase our agency and our ability to act independently within our profession.[8] Nevertheless, because of the heterogeneity in this profession, identifying an archetypal technical communication identity is less important than considering what benefits stronger professional identity can afford.[9] Studies of other professions show that belonging to a professional association can strengthen an individual's sense of belonging, professionalism, and autonomy.[10] The impetus for professionalization of technical communication was driven by professional associations. This drive resulted in successful projects: defining a body of knowledge, identifying the skill set, and developing certification and education programs as a potential route to professionalizing. Even as the focus has moved on from professionalization, professional associations undertake several professionalizing activities: they maintain codes of ethics and other standards that govern ethics and quality in practice. They run conferences; publish journals, magazines, or newsletters; represent members; promote technical communication; and support members through training, professional networks, and information on best practices.

There is some evidence that the dilution of professional identity may be a factor in decreasing membership of professional associations. With fewer members and less funds from membership dues, those associations have less scope to support us in the future. This situation demonstrates how important new voices are in this profession. If you are a student, an early-career practitioner, or a lateral entrant to technical communication, you can make a difference in this profession by joining and contributing to a professional association.

An examination of industry practice requires input from professional associations, academics, and industry practitioners. Stronger ties between academia and industry practice could also lead to an exploration of strategic issues, such as professional identity and internationalization. If you are a leader in a professional association or if you are in a position to influence leadership, you can encourage and develop stronger ties among the various professional associations with a view to embarking on a developmental agenda.

Savage linked professional identity with career paths. He suggested that professional consciousness develops when opportunities exist for "practitioners to make a long-term career in the field" (p. 359).[11] At present, career paths are not clearly defined in technical communication. It is not possible to identify career paths within specific regions and sectors or for specific job titles. That is true in many newer professions because of factors such as work patterns in the knowledge economy.[12] Nevertheless, many studies, including this one, show that the current lack of understanding of career paths, the differences between them, and their intersections, is problematic. We also need to better understand the synergies with and differences from other careers. Research on sectors and environments and the ways they shape work and professional identity is also limited. Likewise, we have a limited understanding of regional and global similarities and variations in practice. These limitations suggest intersecting research directions:

- Projects that map job titles to skills and competencies.
- Projects that examine education and career paths in different regions and sectors.
- Projects that explore practices, job satisfaction, and job opportunities in different regions and sectors.

These projects would benefit students, practitioners, professional associations, teachers, and lateral entrants to technical communication. If you are a researcher or if you are planning to undertake research in technical communication, you could follow one or more of these research threads.

Research of/with/within International Contexts

George Hayhoe's commentary on the research base from a 2006 *Technical Communication* (p. 141)[13] editorial still rings true:

> As we recognize that our profession is becoming global not only in terms of the audiences it serves but also in terms of those who practice it, we should likewise acknowledge that our discipline's research base is no longer adequate.

Although this book included voices of international practitioners, it does not dissect features of international practice. Future research must address this limitation. Many projects and studies are conducted in countries throughout the world which either are fully embedded in technical communication or are germane to our profession and practice. If you are a technical communication researcher based outside of North America, your contributions to the collective understanding and development of technical communication are essential. If you are a researcher based in North America, you have an important mentoring challenge and opportunity: to collaborate and co-author work with researchers in other regions, especially individuals whose first language is not English. Professional associations and journals also have a role to play in encouraging more research from Europe and also from Asia, Australia and New Zealand, and other global regions where technical communication is currently practiced or is emerging.

Practice- and Practitioner-focused Engagement

According to Gherardi, research can enable "the representation of practices as a means to empower practitioners" (p. 210).[14] This outlook encouraged me to attempt to represent technical communication practices, gather input from industry practitioners, and potentially empower technical communicators through their involvement.

If you are a practitioner, you know your practice, engage with communities, and negotiate shifting labor market patterns in ways that we academics understand only in the abstract if at all. My experience (from asking regularly!) is that you want to get involved in research projects and you also want to have input into, and are very good at, strategizing. I have observed, from this and other projects, your keen appetite to contribute. Nevertheless, because you have fewer incentives to conduct research than academics do, you need to be supported. Professional associations can encourage and enable collaborative partnerships involving academics and practitioners.

If you are a student, I hope you see from this book both the exciting potential in your future career and how your contributions can support, develop, and sustain your profession and practice.

If you are a teacher or researcher, I hope this book has encouraged you to explore the possibilities of engaging with industry practitioners and professional associations in setting collective research and strategic agendas. Even if your research focus is not obviously practice-oriented, it is possible and important to seek to involve industry practitioners.

I echo Carolyn Rude, however, in acknowledging that our situation would be equally problematic if research were always "tied" to practice. We need an expansive research agenda that allows theoretical, exploratory, experimental, and value-oriented studies, too. As she (p. 373)[3] explained,

> These observations should not be interpreted to say that all research is or should be about practice in comparable disciplines or in ours. Research should never be entirely tied to practice or we would put unnecessary constraints on inquiry and limit potential discovery. We would also make ourselves vulnerable to the vagaries of practice and define our reason for being in limited terms. However, we would all be surprised if, for example, medical practice and medical research had little to do with each other.

Like Rude, I am convinced, however, that more industry and practice-focused research and strategy will help us to develop and to be recognizable and sustainable. If academics, organizations, and industry practitioners can collaborate actively about matters of shared and practical concern, we can accrue benefits for the whole profession. Will the next few years see a movement toward strategic initiatives that can strengthen our professional identity and practice concerns? That is an open question and one that requires us – whether we are students, teachers, practitioners, or community leaders – to find ways to explore and express our shared goals, values, and vision for our profession and its practice.

References

1 St. Amant, K., & Melonçon, L. (2016). Reflections on research: Examining practitioner perspectives on the state of research in technical communication. *Technical Communication, 63*(4), 346–364.

2 Boettger, R. K., & Friess, E. (2020). Content and authorship patterns in technical communication journals (1996–2017): A quantitative content analysis. *Technical Communication, 67*(3), 4–24.

3 Rude, C. D. (2015). Building identity and community through research. *Journal of Technical Writing and Communication, 45*(4), 366–380. https://doi.org/10.1177/0047281615585753

4 Giammona, B. (2004). The future of technical communication: How innovation, technology, information management, and other forces are shaping the future of the profession. *Technical Communication, 51*(3), 349–366.

5 Boettger, R. K., & Friess, E. (2016). Academics are from Mars, practitioners are from Venus: Analyzing content alignment within technical communication forums. *Technical Communication, 63*(4), 314–327.

6 Bloch, J. (2011). Glorified grammarian or versatile value adder? What internship reports reveal about the professionalization of technical communication. *Technical Communication, 58*(4), 307–327.

7 Nicolini, D. (2012). *Practice theory, work, and organization: An introduction.* Oxford University Press.

8 Wilson, G., & Wolford, R. (2017). The technical communicator as (post-postmodern) discourse worker. *Journal of Business and Technical Communication, 31*(1), 3–29. https://doi.org/10.1177/1050651916667531

9 Slack, J. D. (2003). The technical communicator as author: A critical postscript. In T. Kynell-Hunt and G. J. Savage (Eds.), *Power and legitimacy in technical communication: The historical and contemporary struggle for professional status* (pp. 193–207). Baywood.

10 Guerrieri, R. (2010). Learn, grow, and bloom by joining a professional association. *Nursing2010,40*(5),47–48.https://doi.org/10.1097/01.NURSE.0000371128.38511.0d

11 Savage, G. J. (1999). The process and prospects for professionalizing technical communication. *Journal of Technical Writing and Communication, 29*(4), 355–381. https://doi.org/10.2190/7GFX-A5PC-5P7R-9LHX

12 Jablonski, J. (2005). Seeing technical communication from a career perspective: The implications of career theory for technical communication theory, practice, and curriculum design. *Journal of Business and Technical Communication, 19*(1), 5–41. https://doi.org/10.1177/1050651904269391

13 Hayhoe, G. F. (2006). Needed research in global technical communication. *Technical Communication, 53*(2), 141–142.

14 Gherardi, S. (2012). *How to conduct a practice-based study.* Edward Elgar.

Appendix A
Profession and Practice Resources

In this appendix, you will find details of professional associations, communities, and other resources.

Training Providers

Type of training	Website
Certification	Society for Technical Communication's (STC's) Certified Professional Technical Communicator program: https://www.stc.org/become-cptc-certified/
	Tekom Europe's certification program: https://www.technical-communication.org/technical-writing/tekom-certification
MOOCs (Massive Open Online Courses)	MIT Open Courseware: https://ocw.mit.edu/index.htm
	Coursera: https://www.coursera.org/
	Udemy: https://www.udemy.com/
Web-based training	STC: https://www.stc.org/education/online-courses/
	ACS Distance Education: https://www.acs.edu.au/
	LinkedIn Learning: https://www.linkedin.com/learning/
	tekom Europe TCTrainNet: https://www.technical-writing-training-and-certification.com/

International short course providers

Country	Course provider
The UK	Cherryleaf (UK): https://www.cherryleaf.com/training-courses/
Australia	Engineering Education Australia: https://www.eeaust.com.au/courses/professional-skills
	ATTAR (Advanced Technology Training and Research): https://www.attar.com.au/attar-professional-writing/
	Australian Online Courses: https://www.australianonlinecourses.com.au/
Israel	Cow TC: http://cowtc.com/inhousetc.html
	Our Best Words: http://tichtov.com/training-courses/

Professional Associations: Practice-focused

Association	Website
Society for Technical Communication	Web: http://www.stc.org Blog: https://www.stc.org/notebook/
Tekom Europe, the European Association for Technical Communication	Website: https://www.technical-communication.org/
Institute of Scientific and Technical Communicators	Website: https://www.istc.org.uk
Center for Information-Development Management	Website: https://www.infomanagementcenter.com
Australian Society for Technical Communication	Website: https://www.astc.org.au/
TechCommNZ (New Zealand association)	Website: https://techcomm.nz/
Japan Technical Communicators Association	Website: https://www.jtca.org/en/index.html

Other Practice Communities

Community	Description	Website
TechWhirl	Magazine, discussion forum, recruitment	http://www.techwr-l.com/
Reddit: r/technical writing	Discussion forum	https://www.reddit.com/r/technicalwriting/
Information 4.0 Consortium	Blog, discussion forum, and events	https://information4zero.org/
Write the Docs	Blog, recruitment, discussion forum, and resources	https://www.writethedocs.org/
GitHub	Resource repositories	https://github.com/topics/technical-writing
LinkedIn	Multiple global and local communities: discussions, recruitment, and announcements	https://www.linkedin.com/
Season of Docs	Google volunteer project	https://developers.google.com/season-of-docs
Tech Writers without Borders	Volunteer documentation projects	https://techwriterswithoutborders.org/

Professional Associations: Academic

Association	Location and URL
Association for Computing Machinery (ACM) Special Interest Group in Design of Communication	Website: http://sigdoc.acm.org/
Association of Teachers of Technical Writing	Website: https://attw.org
Council for Programs in Scientific and Technical Communication	Website: https://cptsc.org Blog: https://cptsc.org/announcements/
IEEE Professional Communication Society	Website: https://procomm.ieee.org/

Other Academic Communities

Community	Resources	Website
Women in Technical Communication	Twitter feed, blog, and resources	https://womenintechcomm.org/
Society for Technical Communication academic special interest group	Discussion forum, mentoring, and events	https://www.stc.org/communities/academic-special-interest-group/
International University Network in Technical Communication	Meetings and resources	https://www.technical-communication.org/technical-writing/international-university-network-in-technical-communication

Communities: Corporate

Company	Resources	Website
Cherryleaf	Podcast, blog, and resources	https://www.cherryleaf.com
Scriptorium Publishing	Blog, research, and podcast	https://www.scriptorium.com/
The Content Wrangler	Blog, podcast, webinars, and resources	https://thecontentwrangler.com/
Zoomin Software	Webinars and blog	https://www.zoominsoftware.com/

Conferences

Organizer	Conference title	Locations	Website
Society for Technical Communication (STC)	STC Summit	Various locations in the US	https://summit.stc.org/
Tekom Europe	TC World and other conferences	Stuttgart, Germany and various locations in Europe and Asia	https://www.technical-communication.org/tekom/conferences/conferences-overview
Institute of Scientific and Technical Communicators	Technical Communication UK Conference (TCUK)	The UK	https://istc.org.uk/
Association for Computing Machinery	SIGDOC	Various locations in the US	http://sigdoc.acm.org/conferences/
IEEE Professional Communication Society	ProComm Conference	The US, Canada, Europe	https://procomm.ieee.org/conference/
Council for Programs in Technical and Scientific Communication (CPTSC)	Conference of the CPTSC	Various locations in the US	https://cptsc.org/conference/
Association of Teachers of Technical Writing (ATTW)	Conference of the ATTW	Various locations in the US	https://attw.org/
LavaCon	LavaCon	The US and Europe	https://lavacon.org/
Adobe	DITA World	Online	https://2020-adobe-dita-world.meetus.adobeevents.com/
Center for Information-Development Management	DITA North America, DITA Europe, IDEAS, Best Practices, ConVEx	Various international locations/online	https://www.infomanagementcenter.com/events/cidm-conferences/
MadCap	MadWorld	Europe and the US	https://www.madcapsoftware.com/madworld-conferences/
Write the Docs	Write the Docs Conference	Various international locations/online	https://www.writethedocs.org/conf/

Appendix B
Ethics Committee Documentation

In order to be approved to gather the narrative data, I needed to submit documentation, together with an application form, to the Research Ethics Committee of the Faculty of Arts, Humanities and Social Sciences at the University of Limerick. The documentation comprised the following:

- An information letter.
- A draft survey.
- A recruitment email.

Information Letter

September 2019

Dear participant,

My name is Yvonne Cleary and I am a lecturer in Technical Communication and Instructional Design in the School of English, Irish, and Communication at the University of Limerick, Ireland. I am undertaking research for a book about technical communication practice.

I am gathering narrative first-person retrospective and contemporaneous work histories to build a stronger profile of technical communication practice and to strengthen our collective understanding of the profession. I am inviting you to participate in the study by responding to a survey.

The survey comprises demographic and open-ended questions about your educational and work background, your current practice, and your expectations and plans for the future. Although there are just seven questions, the survey will probably take you about 40 minutes to complete. You should not provide any identifying information (such as your name or the name of the organization you work for) when completing the survey.

If you participate, your responses will be reported in snippets or combined to generate composite representations that will be reported as vignettes. Therefore, you will not be identifiable in any reporting of the data and your anonymity and confidentiality are ensured. Because responses are entirely anonymous, it is not possible to withdraw your responses after you submit the survey.

The study will be published in a book, tentatively entitled *Industry Perspectives on the Profession and Practice of Technical Communication*.

You are not obliged to complete the survey, and you can skip any questions that you do not wish to complete.

If you have any queries or would like more information about the project, please contact:

* Principal investigator: Yvonne Cleary (yvonne.cleary@ul.ie)

This research study has received ethics approval from the Arts, Humanities and Social Sciences (AHSS) Research Ethics Committee (2019-06-08-AHSS). If you have any concerns about this study or your participation and wish to contact an independent authority, you may contact:

Chairperson, AHSS Research Ethics Committee
AHSS Faculty Office
University of Limerick
Tel.: +353 61202286
Email: FAHSSEthics@ul.ie

Best regards,
Yvonne Cleary

Draft Survey

Introduction

Thank you for participating in this survey. I am gathering narrative first-person retrospective and contemporaneous work histories to build a stronger profile of technical communication practice and to strengthen the identity of the profession.

Please be sure not to include any information that could identify you personally in the responses (e.g., your name or the name of the organization you work for).

Your responses will be reported only in snippets or combined to generate composite vignettes. Therefore, you cannot be identified in any reporting of the data and your anonymity and confidentiality are ensured.

The study will be published in a book, tentatively entitled *Industry Perspectives on the Profession and Practice of Technical Communication*.

If you have any questions about this study, contact me at yvonne.cleary@ul.ie.

This research study has received ethics approval from the Arts, Humanities and Social Sciences (AHSS) Research Ethics Committee (2019-06-08-AHSS). If you have any concerns about this study or your participation and wish to contact an independent authority, you may contact:

Chairperson, AHSS Research Ethics Committee.

AHSS Faculty Office
University of Limerick
Tel.: +353 61202286
Email: FAHSSEthics@ul.ie

To proceed to the survey, click on the link below. By submitting the survey responses, you are consenting to the terms outlined above. It is not possible to withdraw your responses after you submit the survey.

Survey Questions

Section 1: Demographic questions

1. What is your gender?
   - Male
   - Female
   - Other
   - Prefer not to say
2. What is your age range?
   - 20–29
   - 30–39
   - 40–49
   - 50–59
   - 60+
3. In what country do you live and work?

Section 2: Narrative questions

1. Describe your work history in your own words. You can refer to any or all of the following:
   - Your education
   - Your first job
   - Relevant training
   - The path that led to your current role
2. Describe, in your own words, a typical work day. If no day is "typical," describe typical activities you undertake. Refer to any of the following features of your work that are typical.
   - Features of your workplace (type of industry)
   - Your role
   - Content you produce
   - Essential software and/or hardware tools and/or applications
   - Professional development activities
   - Collaboration
   - Involvement in networks
3. Describe, in your own words, what changes and developments you predict for your job in the next five years and/or in the longer term.
4. Outline how you will prepare for those changes.

Recruitment Email

Do you work in technical communication or a related discipline?

If so, I would greatly appreciate it if you could complete the linked survey. The survey has only a small number of questions, but they all require detailed responses, so it may take up to 40 minutes to complete. All your contributions are welcome.

Take the Survey
What is this survey for?

My name is Yvonne Cleary and I am a lecturer in Technical Communication and Instructional Design at the University of Limerick, Ireland. I am undertaking research into technical communication practice.

Will my contribution be anonymous?

Yes, your identity will be anonymous. See the attached Information Sheet for further information.

The Information Sheet also contains additional details about the study.

If you have any questions about this research study, do not hesitate to contact me.

Please share this message with anyone in your network who fits the profile and who might be interested in participating.

Many thanks!

Yvonne Cleary,

University of Limerick, Limerick, Ireland

Index

academic developments 25–28
academic identity 25–26
academic journals 12, 26, 43, 58, 63, 67, 117, 216, 230, 232
academic programs: alternatives to 41–42; *see also* professional development; certificate 29–32, 136; content of 35; international 30–31; master's 16, 26, 29, 32–34, 36–41, 49–50; online 35; PhD 29–32; spotlight 37–41; titles 29–32; undergraduate 30, 33, 35–36, 50, 186, 203; value of 28; vignette 33–34
academic qualifications: levels 31–33, 35; working without 25, 27
Acrolinx 97, 106, 161
activities: core 90–102; *habitus* 186; practice theory and 197; range of, 6, 20, 92, 199
adaptability 41, 109, 170, 179, 228
Agile 17, 66, 70, 75, 91, 98–99, 110, 142–143, 178
alumni 56, 72, 211
animation 97, 107, 119–120
API documentation 43, 102, 170, 176–177; *see also* developer documentation
applications *see* tools
artifacts 116–121
artificial intelligence 11, 141, 155, 157
AsciiDoc 106–108, 173
Asia 29–30, 51, 61, 64, 130, 225, 232, 237
Association for Computer Machinery Special Interest Group on Design of Communication (ACM SIGDOC) 62, 64, 69, 236–237
audience analysis 88, 100–101
Australia 4–5, 14, 29–30, 61, 69, 168, 232, 234

Australian Society for Technical Communication 7, 60–61, 235
automation 11, 141, 155–159, 161–162

blogs 5, 58, 60, 68, 85, 87, 94, 103, 126, 129, 131, 156, 191, 215, 217
Blumer, H. 192
book series 60–63
Bourdieu, P. 186–187, 193, 194, 197, 204
Brumberger, E. 87, 89, 129, 131–132, 202–203
Bundesagentur für Arbeit 14
Bureau of Labor Statistics 14, 130, 145–146

Canada 5, 14, 30, 64, 140, 237
career: change 19, 25, 55, 86, 126, 154, 186; early 19–21, 24, 33–34, 69, 136, 230; history 220; long-term 231; management 47, 180; mentoring 179; paths 17, 128, 220, 231; planning 178; second 5, 25, 41, 227; stage 143, 150, 228; trajectory 6, 213
Carliner, S. 44, 109, 143–144
Carroll, J. M. 11
CCM *see* component content management
Center for Information-Development Management (CIDM) 100, 104, 116–117, 237
certification: benefits, 45–46; limitations, 47; providers 234; STC model 30, 46; strategic importance of 46; tekom Europe model 30, 46–47
chatbots 41, 95, 158–159, 175
Cherryleaf 30, 70, 87, 234, 236
China 30, 51, 59, 129–130
Clarke, A. 214
Closs, S. 17, 37

coding 69, 103, 106, 109, 173, 177, 179
Communication Design Quarterly 27, 62
communication skills 88, 90–92
Communicator magazine 66–67
communities of practice 44–45, 55–57, 78, 197–198, 212, 223, 228
communities: academic 71; academic/ practitioner 71; alumni networks 72; blogging 68; challenges of 77–78; definition 55; online 6, 67–73; practitioner 69–71; professional identity and 55–57; project 72; volunteer 70; workplace 73–77
competence: definition 86; framework *see* TecCOMFrame
competencies and skills: durable 169; graduate 41; new and emerging 170; required 88–90; research about 86–87; stretching 177; transversal 49
component content management (CCM) 100, 117–118, 172–173
conferences 44, 60–66, 69, 71, 100, 142, 216, 230, 237
constellations of practice 198
content: analyses 221–222; intelligent 11, 116, 170–174; modular 11, 14, 156, 172; *see also* topic-based writing; personalized 11, 159–160, 163–164, 171–172; reuse 118, 172–174; trustworthiness 169
content development 26, 29, 31, 46, 88, 92–98, 104, 157, 173
content management 11, 35, 46, 61, 100, 105, 107, 149, 173, 216; *see also* component content management
content management system 33, 97, 107, 111
content strategy 15, 29, 31,61, 70, 88, 101, 149, 216
Content Wrangler 70, 172, 236
Coppola, N. 190
Council for Programs in Scientific and Technical Communication (CPTSC) 26–27, 62, 65, 236, 237
Covid-19 5, 133–136, 166–169, 217, 229
creativity 90, 121, 133
critical thinking 32, 89–90
Crowdsource TPC 73
cultural capital 63, 187
curriculum: development 26, 37–41, 203; planning with TecCOMFrame 49–50; typical content 35

Darwin Information Typing Architecture *see* DITA
data: analysis 222–223; big 141, 159–160; gathering 218–222; narrative *see* practice narratives
design thinking 96
developer documentation 8, 69, 102, 108, 174, 176
digitalization 14, 31, 35, 155, 157, 167, 200, 227
digital transformation 8, 134
DITA 11, 43, 65, 95, 103–104, 106, 108, 110, 115–116, 118, 158, 164, 173–174
diversification 14–15, 141, 148, 159–161, 171, 195, 224, 227
diversity: disciplinary 6, 12, 58, 198, 227; of education and training 18–19, 36, 195; inclusion and 203; role 5–6, 14, 67, 83; workplace 126
Docs as Code 65, 107, 176, 202
domain knowledge 88–89, 102, 121

e-learning content 107, 118–119
economic capital 187, 190
educational backgrounds 34, 36, 220, 227–228
educational paths 24, 33–34, 36–37, 188, 191
education programs *see* academic programs
electrification 10, 155, 157
engineering: communication 62, 64; education 10, 25–26, 29–30, 38; sector 17, 32, 102, 128, 130; products 10, 201; teams 75–76
ethical implications: of automation 179; of big data 160
ethics: codes of 57–58, 188, 230; teaching 35; in research 12, 219, 238–241
Europe 4, 10, 26, 29–30, 35, 48–50, 61, 69, 92, 96, 129–130, 132, 135, 148, 160, 211, 216, 221, 232
European Commission 134
European Qualifications Framework 51
European Union 47, 87, 148, 160, 196, 211
events of instruction 119

field, definition 186; *see also* Bourdieu, P.
FrameMaker 43, 104–106, 111
France 5, 26–27, 30, 130

General Data Protection Regulation
(GDPR) 160, 220
genre 6, 13, 35, 88, 112–116, 118, 121,
198, 216, 227
Genres Across Borders 115
Germany 14, 17, 26, 29–30, 37–40, 42, 47,
129–131, 148, 155, 189, 237
Gherardi, S. 199–200, 204, 232
Giammona, B. 190, 229
gig economy 162–163
GitHub 69, 107–108, 177, 235
globalization 18, 26, 133, 138, 140, 154,
164–166, 227
Goffman, E. 193–194, 224
groupthink 78

habitus 186–187, 193–194, 197; *see also*
Bourdieu, P.
Hayhoe, G. F. 42, 89, 222, 231
Henze, B. 86, 112, 114
Hui, A. 197
humanities 26

IEEE Professional Communication
Society 62–64, 236–237
*IEEE Transactions on Professional
Communication* 12, 26, 62–63
India 27, 30, 42, 59, 61, 92, 130,
139, 149, 203
Industry 4.0, 155–156, 171, 179
industry sectors 6, 11, 17–18, 87, 94, 102,
115, 126, 128–132, 140, 145, 148–149,
167, 212, 228, 231
information architecture 39, 61, 65, 88, 101
Information 4.0 Consortium 156, 175
information and communications
technology (ICT) 6, 129, 211, 228
information design 15, 17, 29, 31, 34–35,
88, 92, 96–97, 114, 202
information development *see* content
development
Institute of Scientific and Technical
Communicators (ISTC) 6, 44, 60–61,
64, 66–67, 87, 189, 202, 221,
235, 237
instructional design 6, 12, 14, 16, 31, 119,
128, 169
Intercom 32, 44, 59–60, 129, 140, 143, 164
International University Network in
Technical Communication (IUNTC)
71, 236
internet: access 113, 136; applications 6,
115, 130, 174; development of 11, 25,
155, 161; impact on education 26–27

Internet of Things 11, 155
Ireland 5, 26–27, 30, 70, 129, 139, 187,
191, 203, 211–212, 216
Israel 5, 27, 30, 158, 234
IT Jobs Watch 148–149

Japan 60–61, 130, 235
job advertisements 34, 46, 69, 87, 89, 102,
104–105, 109, 117, 129–130, 148, 203
job opportunities 16, 45, 128, 130,
145–148, 231
job satisfaction 76, 135, 142–145, 228, 231
job titles 5, 8, 14–15, 31, 66, 75, 89, 92,
145–149, 191, 195, 200, 216–217,
227–228, 230–231
Johnson, T. 68, 75, 87, 90, 104, 120, 135,
168, 174, 176, 202
*Journal of Business and Technical
Communication* 12, 26
*Journal of Technical Writing and
Communication* 12

Kunz, L. 100, 102

labor market 5, 14–15, 27–28, 32, 63, 142,
145–149, 164, 167, 232
labor models 67, 154, 162–163, 228
Lanier, C. 133, 140, 143, 173, 202–203
Larson, M. 193
lateral entrants to technical
communication 20, 224, 230–231
Lauer, C. 87, 89, 129, 131–132, 202–203
legitimate peripheral participation 44–45,
73, 198, 201
LinkedIn 44, 56, 61, 68–72, 105, 107, 142,
216
LinkedIn Learning 30, 44
literature review 218–219
localization 3, 16, 63, 70, 95, 165, 176;
see also translation
lone writers 59, 76, 91, 131, 136,
141–143, 176, 191, 202, 212
Lone Writer SIG 76

machine learning 155, 157–159
MadCap Flare 43, 65, 104–106, 108
management: career track 67, 133; of
people 100; project 17, 27, 45, 91,
98–99, 103; quality 99; time 43, 86, 89,
99, 105, 178; *see also* content
management
management competencies 88, 98–100
manuals 11, 17, 26–27, 43, 95, 97, 105,
114–117, 156, 162, 171, 175, 201

Markdown 106–109, 173, 176
marketing communication 17, 40–41, 90, 100, 117, 120, 130, 147–148
massive open online courses (MOOCs) 30, 234
mentoring 33–34, 71, 177, 179, 232, 236
metadata 89, 101, 118, 158, 164, 172–173
microcontent 11, 170, 174–176
Microsoft Teams 104–105, 107, 134, 137–138
Microsoft Word 104–107, 110, 114, 222–223
military 10, 128
Miller, C. R. 54, 113, 192
MOOCs *see* massive open online courses

narrative inquiry 213, 219
national occupation classifications 14
needs assessment 93, 101
New Zealand 14, 27, 29–30, 60–61, 232, 235
Nicolini, D. 197, 199–200
North America 4, 14, 18, 29–30, 35, 48, 51, 59, 62, 65, 92, 130, 201, 203, 221, 229, 231–232, 237

offshoring 133, 138–140, 150, 162, 216, 228
open-source: movement 65, 163; projects 69–70; software 103
outsourcing 6, 133, 138–140, 150, 161–162, 216, 228

pandemic *see* Covid-19
peer review 91, 97–98, 111
personal computer 11, 25–26, 117, 155
personalization 95, 164, 172; *see also* content
perspective: book 19; international 18, 20, 212; industry 4, 10, 218; professional 8–9, 186; symbolic-interactionist 192, 214; user-centered 8–10
PhD *see* academic programs
plain English 94
podcast 43–44, 70, 95, 105, 121, 168, 221, 236
postmodern practice 170, 202–203, 214–215
practice: education intersections 203–204; definitions 197; industry 4, 60, 69, 178, 200, 229–230; international 203, 231; normative conceptions of 199–200; researching 199–200; researching in technical communication 200–204; research tied to 232; theory 197; turn 196; variation in 198–200, 231

practice narratives 4, 33–34, 36, 109–111, 132–133, 136–137, 141–142, 219–221, 223–224
problem solving 86, 89–90, 144, 228
procedural writing 7–8, 94, 112, 174
professional associations: academic 61–63, 236; definition 57; establishing 212; falling memberships 66–67; features and benefits 57–59; lone writers 76; membership of 66, 195, 231; practitioner-focused 59–61; professional development and 43; professional identity and 56, 66–67, 230; related fields 61; resources 87
professional development (PD): continuing 66; examples 43–44, 109–111; informal learning and 43; opportunities 5, 59, 86, 100, 144, 228; planning 34, 153, 177–178; programs 30; research about 191, 198; self-employed 43
professional identity: ambiguous 228–230; career paths and 231; communities and 54–57; composite 194; factors that contribute to 28, 60, 131, 197; factors that dilute 66, 77–78, 195; fragmenting 18, 161; performance and 193–194; professionalism and 192–195; professionalization and 196; social justice and 203; stifling 193; strengthening 230, 233
professionalism 26, 189, 191–195, 229
professionalization: aspirations 28; golden age 188, 191; outcomes of 195–196; problems with 191, 193; process 188–189; research 28, 142, 212, 229; technical communication 189–191; theory 188, 212
professional outlook 125, 142–149
profession and practice intersections 186, 212, 218
professions: definition 188; diversification of 58, 141, 164, 227; future of *see* work, future of; globalized 221; negative traits 189; new and emerging 28, 154, 159, 166, 202, 231; prestige 187, 189; regulated 35; traditional 165–166, 188, 190–191, 195; traits of 188–193, 195
public sector 128–129, 166
publishing formats 116–117

Reddit 69
remote learning 169

remote working: advantages 134–135; challenges 135–136; Covid-19 and 134–135, 148, 168; variables 136; vignette 136–137

Rensselaer Polytechnic Institute 25–26

research: academic/industry divide 13, 21, 67, 71, 149, 196, 203, 216; agenda 229, 232; "bigger picture" 199, 204; challenges 13–14, 228; international 14, 18, 149, 203, 225, 231–232; limitations 217, 219, 221, 224–225; skills 93–94; strands 13, 218; strategic direction of 228–233; qualitative 143, 168, 199, 211, 213, 217, 219–220, 222; symbolic interactionist 192, 199, 213–214; with information users 13, 93–94

research questions: global 13; for this study 217–218

restructuring, corporate 133, 140–142, 165–166, 228

reusable content *see* content, reuse

rhetoric: composition and 5, 9, 12, 29, 34, 63, 71; intercultural 27; research 12–13

rhetorical situation 112

rhetorical skills 88, 94, 109, 170, 228

routinization 157, 161

Rude, C. 13, 204, 218, 232

Savage, G. J. 28, 46, 190, 201, 231

Schatzki, T. 197, 200

Schema.org 174

Schriver, K. A. 7, 13, 96, 100–101

screencasts 120

Scriptorium 70, 87, 101, 120, 236

scrum 91, 98–99, 132, 143

Season of Docs 70, 235

second-language skills 95–96

Second World War 10, 26

self-service platforms 11, 158, 164, 168–169, 228

silos 156

Simplified Technical English 95

single sourcing 11, 118, 172–173

situational analysis 210, 214–218

situational map 215–218

skills: definition 86; research about 86; required 87–88; technical 103–108; transferable 89; writing 30, 94–96, 178

social capital 187

social justice 12–13, 35, 127, 203, 208

social media content 95, 117, 120–121

Society for Technical Communication (STC) 6–7, 27–30, 44, 46–47, 59, 61, 63–64, 66, 71–72, 76, 142, 196, 200, 202–203, 234–236

software development 17, 74–75, 98, 102, 149, 201

Spilka, R. 11, 200

St. Amant, K. 91, 168, 230

STC Summit 28, 64, 66, 237

strategic agenda 228–232

structural functionalism 188, 192

style guide 33, 76, 95, 97, 106, 110–111, 117, 142, 195

subject-matter experts (SMEs) 17, 33, 74, 91, 93, 97, 102, 110–111, 138, 141

survey instrument: benefits 220–221; design 220; limitations 221; purpose 4, 219

Susskind, R. & Susskind, D. 154–166

Sweden 5, 26, 30–31, 36

symbolic capital 187–188

symbolic interactionism *see* research

teams: Agile 98–99, 142–143; configurations 74; cross-functional 74–75, 195; documentation 75; virtual 136–138

TecCOMFrame 27, 31, 47–50, 72, 87, 99, 103, 196, 216

Tech Writers without Borders 70, 235

Technical Communication 12, 26, 42, 59, 62, 190

technical communication: academic discipline of 10, 18, 25–27, 51; boundaries 5; definitions 7–8; distinctiveness 6, 15, 18, 21, 100, 169, 199, 224, 227; historical roots 9–11; profession 4, 10, 14, 16, 18–19, 145, 179, 189–191, 201, 215; recognition for 14, 145; research base 12–13; theoretical base 9, 12; umbrella label 5, 14, 16, 227

Technical Communication Body of Knowledge (TCBOK) 7, 27, 47–48, 50, 59–60, 72, 87, 190, 196

technical communication practice *see* practice

Technical Communication Quarterly 12, 26, 62

Technical Writers of India 61

TechWhirl 25, 69, 128, 133, 235

tekom 27, 59, 87, 189

tekom Europe 6–7, 29–30, 42, 44, 46–50, 59–64, 71–72, 79, 129, 131, 196, 221, 234–235, 237

template 114

theory: connection to research 9, 12, 183; definition 9; grounded 213–214, 222; rhetorical 96; *see also* practice; professionalization

TikTok 171
time line: educational development 25–27; professional development 9–11
Tina the Technical Writer 201
tools: commonly used 105–107; functions 105–107; industry-standard 104, 106, 108; limitations 108–109; open-source 103; range of 103–104, 119; selecting 8, 57, 88, 173
topic-based: formats 11, 117–118; writing 35, 95, 105, 115, 118, 172–173
trade publications 4, 59–61, 103, 126, 215, 218, 221
traditional professions *see* professions
training *see* professional development
transferable skills 89–90
translation 16, 21, 35–37, 39, 48, 95, 107, 118, 129, 165; *see also* localization
Twitter 44, 68, 71, 105, 107, 120, 216, 236

United Kingdom (UK) 10, 29–30, 36, 42, 61, 64, 66–67, 70, 87, 130, 138, 148, 189, 202–203, 234, 237
United States (US) 10, 13–14, 18, 25–26, 29–30, 35–36, 50, 59, 64–65, 69, 112, 128–130, 138, 140, 145–146, 168, 202–203, 216, 228–229, 237
University of Applied Sciences, Karlsruhe 17, 30–31, 37–41
University of Limerick 16, 31, 72, 126–127, 211, 238–241
user assistance 15, 39, 94–95, 116–119, 147, 161–162
user assistance, embedded 11, 103, 106, 116, 118, 121, 155–156, 162, 174–175
user experience (UX) 12, 16–17, 33–34, 39, 59, 63–64, 75, 88, 93–94, 96, 101, 117, 147–148, 174–175
user manual *see* manuals
user research *see* research

user-centered design 7–8, 31, 94
user-generated content 11, 119, 162–163, 171
UX design 35, 175
UX writing 15, 33–34, 147, 175–176

video content 96, 120–121, 171
vigenttes: choosing a program 33–35; composing 223–224; content writer 132–133; contractor 136–137; lone writer 141–142; purpose 4, 19–20, 223; typical day 109–111
virtual teams *see* teams
visual design *see* information design

webinars 33, 43–44, 59–60, 66, 70, 111, 168, 172, 221, 236
wikis 11, 106, 116, 156
Wenger, E. 44, 56–57, 197–199, 223
work environments 131–132
work, future of: demystification 166; emerging skills 159–160; end of an era 155–156; more options for recipients 163–164; new labor models 162–163; patterns 154–155; preoccupations of professional firms 164–166; preparing for 177–179; professional work reconfigured 161–162; transformation by technology 156–159
working from home *see* remote working
workplace: cultures 126–128, 131; sociability 76–77, 137; studies 202–203, 212
Write the Docs 65, 69, 107, 168, 235, 237

XML 43, 103–104, 106, 108, 110, 118, 173

YouTube 94, 119–120, 171, 201

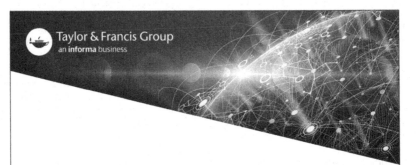

Taylor & Francis eBooks

www.taylorfrancis.com

A single destination for eBooks from Taylor & Francis
with increased functionality and an improved user
experience to meet the needs of our customers.

90,000+ eBooks of award-winning academic content in
Humanities, Social Science, Science, Technology, Engineering,
and Medical written by a global network of editors and authors.

TAYLOR & FRANCIS EBOOKS OFFERS:

A streamlined
experience for
our library
customers

A single point
of discovery
for all of our
eBook content

Improved
search and
discovery of
content at both
book and
chapter level

REQUEST A FREE TRIAL
support@taylorfrancis.com

Printed in the United States
by Baker & Taylor Publisher Services